GW01376845

Print Buying

Also by Roy Brewer
AN APPROACH TO PRINT (*Blandford*)
ERIC GILL, THE MAN WHO LOVED LETTERS (*Muller*)

Print Buying

Roy Brewer

DAVID & CHARLES
Newton Abbot   London   North Pomfret (Vt)

British Library Cataloguing in Publication Data

Brewer, Roy, 1924–
   Print buying.
   1. Printing – Technological innovations
   I. Title
   686.2   Z249

   ISBN 0-7153-8554-2

© Roy Brewer 1986

All rights reserved. No part of this
publication may be reproduced, stored
in a retrieval system, or transmitted,
in any form or by any means, electronic,
mechanical, photocopying, recording or
otherwise, without the prior permission
of David & Charles Publishers plc

Typeset by Typesetters (Birmingham) Ltd
Smethwick, West Midlands
and printed in Great Britain
by Butler & Tanner Ltd, Frome and London,
for David & Charles Publishers plc
Brunel House   Newton Abbot   Devon

Published in the United States of America
by David & Charles Inc
North Pomfret   Vermont 05053   USA

Contents

Introduction		7
PART ONE: The Print Buyer's Role		11
1	The changing role of the print buyer	13
2	What technology?	15
3	Back to basics	18
4	Using the budget	28
5	Selecting suppliers	35
6	Working with designers	39
7	In-plant printing	42
8	Computerised graphics	52
9	Colour	61
10	Presswork	75
11	Binding and finishing	87
PART TWO: Technological Changes		93
12	Copiers	97
13	Ink jet printing	100
14	Electronic printers	102
15	Laser printers	105
16	Facsimile transmission	108
17	Raster image processing	111
18	Microrecords	114
19	Electronic publishing	116
20	Beyond the 80s	119
Index		123

Introduction

Print buying is a job which requires developed skills. The print buyer's role is hard to define, but broadly it involves the planning, preparation, budgeting, costing, ordering, control and monitoring of print production. How that is done will differ from place to place, in the resources available to do it and in the kinds of print bought. Today it will almost certainly extend to purchasing services ancillary to print, such as typesetting, colour reproduction and specialised finishing, and some origination such as photography and artwork. The field is wide but some techniques are basic. It is these, and the technical knowledge which goes with them, with which this book is mainly concerned.

I am well aware that there are women print buyers, but refuse to fall into the him/her-he/she trap.

Print buying is the management of *all* the resources available to meet an organisation's print needs. Because those needs differ it is not possible to frame a comprehensive set of rules which any print buyer can apply; but it is increasingly necessary for him to make choices, judgements and decisions about the many routes which print can now take from inception to completion and here guidance can be given. Due to technological and other changes in and around print, the print buyer is likely to get less help than, at one time, he could expect from his suppliers in a neatly circumscribed 'printing industry'. The new systems will be as unfamiliar to some of them as they may be to him.

Overall there is now a greater range of specialisations which affects the ways a print job can be planned and executed. This makes print buying itself more specialised and technologically orientated. Some of the technology lies outside the printing industry altogether, and can involve the use of word processors, copiers, facsimile transmission systems, business computers and other equipment now available to commerce and industry. The print buyer needs to know how it can affect the way he works with printers.

I have tried to keep jargon to a minimum, but assumed a basic knowledge of printing, and of data processing which is central to many of the newer systems and their applications. Throughout I have endeavoured to concentrate on what things do rather than on how they work.

There is no simple, clear division between 'new' and 'old' technology in print. Some services are no longer available, and others are available only from new sources using new equipment; but much of what is carried out by the latest systems and methods ends up looking the same as it did when printed by conventional ones. Most of the new technology is concentrated in origination and pre-press areas. The 'input' may be different but the 'output' – print – usually looks familiar. The print buyer still needs many of the same skills he used for planning, purchasing and progressing a 'traditional' job. His role – as the first chapter suggests – may have changed, but his objectives remain those of using available resources efficiently and economically.

Finally, it could well be that the title 'print buyer' is becoming obsolete, at least as it has conventionally been used. Print will continue to be needed, and will remain an important part of the expenditure of any organisation which uses it in volume; but it is possible – indeed probable – that dealing with printers will be only part of the job. The print buyer of yesterday could well be the production manager or 'reprographic' manager of today, taking in a broader spectrum of less clearly defined activities. More people will be involved in meeting an organisation's communications needs than there were in the immediate past, when the print purchasing function predominated; but whatever title is used, somebody has to co-ordinate things and print on paper is still the best, cheapest and most versatile medium for a vast range of needs.

It is sometimes forgotten that the media which are said to be 'replacing' print are themselves highly dependent on it. Changes have nevertheless taken place for printers, as for their customers, in many of the products required and the means of supplying them in faster-moving, more competitive and more precisely targeted markets.

So I shall continue to address the 'print buyer' in the full knowledge that he might not be designated as such by his own organisation, and could have other responsibilities which are part of a wider 'communications industry'; but where print is required print must be bought, and that is still no easy thing to do efficiently.

Part One

The Print Buyer's Role

1

The changing role of the print buyer

Buying print is buying services and materials, some of which will be outside the printing plant and not produced by the printer. All these resources must be selected and managed. In the past it was usually possible to use a small number of known services and suppliers, work within a fixed print budget and place orders routinely. It is now harder to do this for the majority of printed items because production methods have changed and, as a result, the printing industry – the print buyer's marketplace – has changed. A print buyer now has to be more observant and more selective to get what he wants. He is not only 'shopping for print' but also evaluating various methods of progressing a job and choosing between them.

Just as technological changes influence the ways a job can be done they have also influenced the structure of the printing industry itself in two important ways. One is in an increased specialisation of production equipment and, therefore, of suppliers, and the other a dispersal of services, which makes it harder to find everything needed to see a job through from a single supplier and in one place. The balancing of production resources is not made easier for the buyer by changes which can take place in the ways work is originated: text and artwork can now take unfamiliar forms. It may originate from computer databases, from word processors or from the electronic aids available to designers, typographers and photographers.

THE CHANGING ROLE OF THE PRINT BUYER

In place of jobbing printers, which have largely disappeared from the commercial printing scene, there are franchised chains of print-shops offering a retail service for copying, printing and typesetting, many new trade services and a growth in captive 'in-plant' printrooms maintained by many organisations to meet part of their own print needs. These, too, cannot be ignored by the buyer as sources of supply.

The practical result of these developments is that almost all but the simplest jobs need careful planning, detailed specification and careful synchronisation in order to come together at the right place, at the right time and as specified. More people are involved than there were when contracts were made mainly between buyers and printers. The buyer must know, or discover for himself, where to look for what he wants.

It is easy to assume that the challenge is simply one of understanding and using the appropriate technology to get the required results. It is not only that. If the buyer understands how the new equipment works and what it does, he will still have to ensure that any part of a job which is produced in one place can be progressed in another. It is, for example, useless to expect a job which has been planned and originated to a fixed size to be produced on a web-offset press which is not able to print it in that size. One of the most frustrating and, in many cases, self-defeating aspects of 'new technology' is that it has developed piecemeal: a lot of it works only as 'systems' which are interdependent and thus prevent the work from being 'portable' between one system and another.

The print buyer is, therefore, not only looking for the technology, but also for the *appropriate* technology to carry out the job in hand.

2

What technology?

A good deal of attention has been paid to 'new technology' in printing; but one thing is certain. It cannot be applied in isolation from what I suppose we shall now need to call the 'old' technology. Virtually all changes of significance to the printer and his customers have taken place in pre-press areas such as typesetting, colour separating, page planning and makeup. There have been developments on presses too but, apart from an increase in mechanical speeds, there has been little fundamental change in the ways they print.

To make the best use of the equipment in a printer's plant the buyer must not only understand its function but he must also evaluate it in terms of what it does for the jobs he wants it to do. His task is not made easier by the reluctance of many printers to explain or demonstrate how the best uses can be made of new systems. For example it has taken twenty-five years or so for computerised phototypesetting to yield substantial benefits. It was not difficult to see where its value could lie in text processing for offset litho, which is a photomechanical process, by increasing the speed at which large volumes of text could be set, corrected and printed; but when the computer-driven phototypesetters appeared their performance brought few discernible benefits, either in terms of profitable operation to the printers who used them or in cost, convenience or quality to most of their customers: such systems were, quite definitely, 'horses for courses' but few printers said as much. Only over the last few years has phototypesetting developed to the point at which it can cope adequately with the demands made on it by a wide range of print needs.

If equipment is compatible, it can be used in several different places, with the output of one system being transported or transmitted to another. Printing requires a strict *sequence* of operations be followed to reach the final product – the print itself. Electronic systems can impose stringent preconditions for successful application. They may increase the number of different routes which can be taken to get a job done, and the choice of appropriate routes is not always easy.

The usual effect of changing technology is to feed back some of the printer's traditional tasks to the buyer, or disperse them to other suppliers. Thus a decision to use an office word processor to originate text for phototypesetting changes the route from origination to typesetting; it affects not only where and how the text is generated, but also other parts of the sequence of operations which start with the original keyboarding.

A print buyer may still say he is concerned only with getting what he pays for in the final product. This is sensible only when the whole of the technology used to manufacture a printed product is retained by the customer's suppliers. As soon as any of it moves elsewhere the system expands and also operates differently. It is important to find out how.

Changing technology in the printing plant once required customers to adapt to it in various ways to obtain whatever benefits it was supposed to provide. Today the printer's machinery and equipment is likely to offer flexibility only to those who know how to buy its products. Electronic colour scanners, for example, offer more sensitive control of colour values and better opportunities for the creative uses of colour than do camera-based systems. The challenge for the print buyer is to get to know, and use, this flexibility. Learning what colour scanners are for and how they work is one thing. Discovering what they can be made to do is another. The buyer who 'never goes near a printer's plant' is unlikely to find out either; the best way of evaluating modern printing technology is to see it in action.

For any new, or changed, system the print buyer needs to find out (a) what the equipment is there for, (b) whether it will call for changes in the way orders are planned, prepared and progressed, (c) what benefits can be expected from its use, (d) what changes, if any, need to be made in planning and scheduling, or other procedures such as proofreading, colour correction, author's corrections, etc. It is also wise to examine the output of the system and obtain specimens of work which has been done when using it.

Later I shall be dealing in some detail with the uses to which office technology can be put. If, for example, a word processor is used for text preparation (which would mean that the printer could be supplied with clean copy ready to go straight to press via a phototypesetter) the customer has contributed materially to simplifying the printer's tasks and should look for some tangible benefit for having done so, either in a faster turnaround of the work so processed by the customer or in the total cost of a job. No technology is of value until it is applied.

3

Back to basics

This section goes back to some of the basics which underlie the ways in which print is planned, bought, produced and delivered. Many are broadly the same as they have always been. Some are different. It is, for example, difficult to buy short-run print locally from what used to be called 'jobbing printers', because jobbing printers are now thin on the ground: they have been replaced in large numbers by in-house printing plants, or by franchised chains of 'instant' printshops.

Additionally we need to distinguish between 'commerical printing' and other methods of disseminating information without printing it, for example by using copying machines, computer printout and non-printed communications media such as data banks, viewdata and microfilm.

They need to be taken into the print buyer's consideration if they are available. Some jobs could not be done properly by these alternatives to commercial printing: nobody would attempt to get a mail order catalogue with colour illustrations produced on a copier, or by a small in-house printroom. Commercial printing is still often the most efficient, cheapest and frequently the only way of getting what's wanted.

Scheduling

Scheduling print is not the same as planning it. The print buyer rarely has complete control over the whole plan; but though his responsibilities may appear to start only when the job is ready to be printed, he would be wise to become involved as early as possible in planning and origination of copy or artwork. It is not feasible even to outline a schedule until it is known what has to be printed, in what form the material will be presented and used and when it is wanted.

Before every print schedule, contact should be established with those responsible for the content and appearance of the finished job. These might include copywriters, typographers, artists and others. Specialists tend to work in watertight compartments and under differing pressures. It can sometimes be useful to work out a timetable for the job and circulate it to all concerned. Such a timetable need not be inflexible to start with, but the moment will come when it *must* be applied if the job is to be scheduled. The assistance and advice the buyer can give at this stage will depend on the kind of work being prepared, the experience, abilities and location of those who do it and, not least, the personality of the print buyer himself. If he is seen only as 'the one who deals with the printer' his deadlines may be regarded as unwarranted interference with the way other people do their jobs. Too bad! The existence of deadlines tells people how much time they have to do what's required of them. In the event of delays or changes of plan, the production schedule can be amended with the minimum disturbance. This is particularly important when outside services – typographers, secretarial agencies, word processing bureaux and the like – are used. More schedules come to grief because somebody wasn't told of a change which affects them than from incompetence.

The print schedule – now more properly regarded as a production and delivery schedule – should include all the operations which take place from the time the job is passed to the printer and its delivery to where it is wanted after completion, including such things as proofreading, colour correction, transporting the print and, when trade services are used, the time taken to carry out the work and move the job from one place to another. This element is nowadays more likely to carry weight since more and more printers are limiting their in-house services and they, or their customers, are frequently using trade services for such things as typesetting, colour separation and finishing.

The planning and scheduling of a single item of print can therefore be something of a headache to organise. I have an A4 booklet of 120 pages plus cover on my desk, printed web-offset in four colours for the English Tourist Board. The following were involved, individually and collectively, in its production: an editor, an assistant editor, four editorial contributors, an outside design company, eight illustrators, six photographers, a photographic library, a cartographer and an advertisement department. The booklet was typeset by one firm and printed and bound by another. It was despatched by post to some destinations and by bulk transport to others. Someone has to conduct this kind of orchestra, and see the whole thing into print and into the hands of those for whom it is intended. Information must be given, received and acted upon: when is a reminder appropriate if things are getting behindhand? How, if they are, or if unanticipated changes are made in the original specification, will that affect the schedule?

One of the penalties we are likely to pay for the speed of modern production equipment is a limitation on the number of 'second thoughts' about the content of a piece of print once it is already in a production sequence. If extensive revisions are made after the order has been placed and proofs submitted, there will be added costs not covered in the original budget. The ideal, therefore, is to bring every major item of print as close as possible to a ready state before any production processes are set in train.

No schedule is realistic unless it takes into account the way people work as well as what they produce. Designers and copywriters need 'thinking time' to develop ideas; suppliers have to schedule the loading of machinery and equipment used to produce the work to meet the buyer's delivery times.

Every job which becomes a 'rush job' contains the seeds of crisis or disaster. If the schedule is complete and realistic – really complete, and not leaving out such details as the need to order envelopes and fill them – it will not run like clockwork unless the buyer has allowed for any predictable disturbances that the schedule could suffer on the production side.

High-speed equipment produces high-speed results only if it is used at optimum throughput. Processing times may look spectacularly short but, to be profitable, the workflow to such systems has to be regulated. The buyer is attracting the possibility of additional costs or delays if his job does not reach a high-speed press and bindery line when it is scheduled to do so. The buyer must try to avoid the 'knock on' effects of delayed schedules or pay the price they can cause in time or money. Print which cannot be processed because it has 'lost its place in the queue' may have to be warehoused until it finds capacity, and few printers are now prepared to do this for nothing if the job is a sizeable one: a quantity of print takes up a lot of room. The buyer who counts on a supplier's 'spare capacity' to get him out of such difficulties with work to strict delivery dates is treading a risky path.

Specifying print

Everything which needs printing starts with a specification. This might be no more than a verbal instruction or a rough draft. The fuller and more detailed a specification is the better chance the job has of being done to the buyer's satisfaction and the easier it will be to obtain a reliable estimate of what it will cost.

A full specification is rarely taken for granted by printers, even for expensive work. The conscientious ones do their best to fill in the gaps so that an accurate costing can be made for the customer. Others do not, and there can be nasty surprises when the bill arrives. Materials waste can often be avoided at the specification stage.

Before any specification can be made the buyer needs three pieces of information: what the proposed purpose of the item is, what resources are available for its production and what proportion of his budget is available to get it done. The buyer may already know from experience the answers to all three. If not, he had better find them out. When printers used a well-defined set of craft skills on a more limited range of equipment and products, specification was easier because there was less choice in how a given job could be processed and all production was normally carried out in the same plant. This is now less likely to be the case.

The customer is buying two things: materials and the use of the supplier's plant and skills in converting those materials. The customer may specify how both these things are done, or leave some of the decisions to the printer. Certainly he cannot leave *everything* to his suppliers. A specification should start with as precise a definition as possible of what the printer is required to produce and a time schedule (derived from the buyer's own schedule) for completing the job. This applies not only to the job specification, but also to any request for an estimate. If anything is left out – the paper, for example, or any finishing or despatch instructions – the printer will probably ignore their absence and quote only on what he sees. The more information he has the more accurate will be his quotation.

If a job is being submitted to more than one printer to obtain comparative quotes, the specification must be the same for all, otherwise realistic cost comparisons cannot be made.

This is more easily done for repeat orders, where a specimen is, in effect, a specification; but only if what is being quoted for and what is required are identical, and will be produced using the same grades of materials to the same size and design as the original. The buyer must be sure of this. It is possible to estimate the effect of a changed specification by asking for a repeat order to be costed 'as is' by the printer, and for a second estimate incorporating the proposed changes, which allows a comparative costing to be made.

For example it might be considered worthwhile to 'proof once only' if the buyer is confident that clean copy, artwork, etc can be supplied and a proof is needed only to check that the work has been fully and correctly carried out. If this is done the printer should base his estimate on not having to proof several times and not anticipating heavy author's corrections, and the job should be cheaper by just that amount.

Now that material costs are high the buyer will seek, where possible, to mitigate them by looking for savings in this area, such as using a cheaper paper or binding, or reducing the number of finishing operations. Comparative costing with changed specifications is a skill which buyers can acquire. If they get into the habit of asking for 'the same again' they will become victims of every increase in materials and production costs which is passed on to the customer. The normal way of keeping track of how jobs are done, and what they cost, is to file one specimen of every completed job together with its original estimate, its specification and the printer's invoice.

It is common for buyers to collect and file specimens of work which approximate the kind of print jobs which might be needed in the future. This squirrelling away of other people's print can become something of an obsession. To be of any use it has to be done methodically. Such collections have their uses, but not mainly in specifying print. It is easy to pass a specimen to a printer and ask for 'something along these lines', but that is not a specification unless the job is identical in all important respects, which it probably isn't.

The value of a standard specification form to both buyer and printer is high. Such a form can be structured to cover the kind of work which passes through the print buyer's hands, but should include as much detail as possible even if some of it is not required for all jobs. Some buyers may need more than one form if they are dealing with, for example, internal design or photographic departments or in-house printrooms (see illustration overleaf).

Date. Name of job. Reference number for job. Name(s) of suppliers.

Time schedule
(a) Date at which all copy/artwork/illustrations are required
(b) Date at which first proofs of text/illustrations are required
(c) Date at which all proofs should be finally passed for printing
(d) Delivery date by printer (or despatch date if the printer is despatching).

Typographical specifications
NOTE: These can vary enormously. The minimum would be:
(a) Headline or display types (as marked by designer on layout)
(b) Typeface(s) (Names and point sizes) for headline and text
(c) Measure(s)
(d) Inter-line spacing (leading)
(e) Additional typographical instructions relating to style, eg indentation of paragraphs, use of caps/small caps, letterspacing, folio numbering, etc.

Illustrations
(a) Number and titles (or key)
(b) Nature of originals (eg colour transparencies, black and white photographs, finished artwork, etc)
(c) Date and destination of all material to be returned.

Materials
(a) Page/sheet size
(b) Paper/paper weight
(c) Estimated number of pages

An example of a fairly comprehensive form, though there are many variations

(d) Trimmed size/bleed
(e) Number of pages (estimated)
(f) Number of folds/folding scheme
(g) Other materials, eg covers, required for printing.

Finishing and outwork
This list should include all operations needed to complete the job, eg numbering, perforating, gluing, coding, etc. If trade services are being used and separately specified a delivery date from the trade house to the printer or customer should be inserted.

Costing
It is open to question whether a specification need, or indeed should, contain any specific information on costs. This is an area which requires separate attention and, probably, its own records. However a specification might include the main cost areas as estimated by the supplier as a standing record of what they are supposed to be. If costings are included in a specification there should be room for two entries – one the originally estimated cost and the other the invoiced cost of the job.

(a) Production charges
(b) Delivery charges
(c) Cost of proofs
(d) Run on price
(e) Materials costs
(f) Outwork costs.

Reproduction
(a) Scaling instructions
(b) Special requirements (eg tint laying, cut-outs, vignettes, etc).

Delivery and/or storage instructions

The print buyer should never assume that people 'ought' to know what to do without being told: if a job can be done in more than one way there is a law which dictates that it will be done in the wrong way. If, for example, use is made of trade services, or of internal resources, a job might well have to take an indirect route to the presses, and beyond them. Synchronisation and co-ordination is called for. This has been mentioned in the section on scheduling; it relates to specification when it comes to deciding *how* to put the job into production. If the printer is using trade services it can be safely left to him to instruct them in accordance with the buyer's full specification. If the buyer purchases such services directly it is he who must specify what's wanted, and not forget to tell the printer what he has decided.

A common example of what can happen if the buyer has not told the printer what he proposes doing is for work which has to be passed, after printing, to a trade binder or finisher, to have been produced in a way which makes it difficult (and, therefore, more costly), or even impossible, to produce as specified. If the typesetting does not fit the page layout, or the margins left on the pages do not allow enough space for perforating, or for spiral or comb binding, the job comes to grief, or its specification has to be changed at the last minute. If, when using word processors for direct input to phototypesetters in ways which I shall be describing later, the operator is not instructed about typography and format, the anticipated benefits will not accrue.

To specify something doesn't mean that this is how it will be done unless everybody involved knows exactly what's wanted. An extension of the specification, or maybe a precondition for it, is the briefing of those progressing the job. The print buyer may well be the best, if not the only, person to do this: he is, after all, supposed to know more about the specification and schedule than other people, and it is his reputation and budget which are on the line if things go wrong. It all has to be paid for, and the penalties for neglect or incompetence are likely to be high . . . and conspicuous.

The print buyer works in a time-frame defined by what takes place from the moment a job is defined to the moment it is delivered as specified. If he tries to operate outside that time-frame he risks delays and disappointments. The solution is to anticipate what people need to know and make sure they are instructed. The trade house which does the colour separations or the trade phototypesetter who sets the text might not need – and could be confused by – the full print specification. What, however, everybody should get are clear instructions on what is needed to ensure that their part of it is carried out correctly *and* in accordance with the demands of subsequent operations. The same goes for the origination and creative sides. It's not a matter of telling other people how to do their jobs but of making sure that they know what their jobs contribute to the end result.

4

Using the budget

Print is normally bought to a budget which may, or may not, be negotiable by the print buyer. If it is long-term and inflexible the constraints on him will be many. When materials and production costs rise, as they have been doing rapidly for some time, he has little alternative but to find ways of saving money, usually to the detriment of quality. This assumes that quality standards have been set. If they have not it is likely that some of the budget was already being squandered by buying more than was needed for the organisation's print needs to be met efficiently and economically. Standards are not the same for all jobs and, in the past, it was common to buy print without carefully defining an acceptable standard for a particular item. Now that materials costs have risen, together with preparation and pre-press costs, the actual printing will represent a relatively small proportion of the total price of a job, and specifications are often trimmed in pre-press areas, such as specifying cheaper papers, dispensing with design services or reducing run lengths.

These controls may or may not have the desired result. If, for example, a cheaper paper is used it could create problems on the press or in finishing operations which leads to materials waste; if run lengths are reduced to avoid the cost of storing print, the cost of more frequent reprinting can be higher than the anticipated saving.

If the print budget comes under the control of a purchasing department, of accountants or of middle management it may well be administered by simple arithmetic without knowing

USING THE BUDGET

what hidden waste or additional expense could be involved over the longer term. At the very least, whoever places orders for print must be able to identify cost-areas and co-ordinate production. Without co-ordination a print budget rapidly becomes disorientated. If a manager decides that a new and elaborate brochure will help sales and can put its production in train without reference to the print budget, the print buyer, not the manager, will be left to explain any budgetary shortfall! This is what I mean by the 'management' of print in addition to buying it.

The base cost of any quantity of finished print comprises materials and conversion. To this must be added preparation cost such as design, artwork, colour separation, typesetting (if it is not carried out by the printer) and any external costs such as envelopes, wrappers or postage, which the job attracts.

Today a budget might need to incorporate cost areas where equipment and resources are used outside the commercial print sector, but within the broader communications needs of an organisation. Computers, copiers, word processors and in-house services, including in-plant printing, are all distinct cost-centres which should be included in a 'print' budget and carefully apportioned. None of these internal facilities is free just because it happens to be available.

At one time most graphic communications were handled in one place – the printer's plant. This is now unlikely, due in part to the wider range of equipment at the places where it is originated, and in part to the increased specialisation in the commercial printing industry which leads to a wide dispersion of suppliers. Another cost-area resulting from this dispersion is the transportation of finished or unfinished print from one place to another, and its despatch when completed. If handling, freight, despatch and postage costs are not included as part of a print budget the budget is, for practical purposes, unrealistic.

The best cost controls are applied at the initial planning stage. If the plan is reasonably complete it is easier to translate into full specifications and allocation of costs. Heavy corrections or alterations can often be avoided if the print buyer is involved from the outset. Jobs still go to printers which could be done on copiers because someone has not considered the alternative; some people cannot visualise print on paper until they see it there, then start making radical changes so that a new start has to be made from first, or even second, proof stage when production costs have already been incurred.

The business computer can be useful in planning and controlling a print budget. Printers themselves are increasingly putting these machines to work for costing and estimating but, so far as I know, print buyers do not use them much. The basic costing and estimating programs are cheap and, with a little skill and experience, allow estimated costs to be compared. A typical small computer estimating program will provide all, or most, of the following information: analysis of cost centres for a job; cost of materials; cost of outwork; added value (the printer's profit); run lengths; job specification (size, format, colours, proofing, etc). Not all these data may be immediately available, and not all of them will be completely settled before the order is placed; but it will almost certainly be more informative to have them on hand compared with a few figures scribbled on the back of an envelope.

The interpretation of printers' estimates is an art in itself. Printers will quote only for what the buyer asks, not what he might have in mind; so if there are variables, such as the possibility of differing lengths of run or changes in the colour content of pages, a request has to be made for alternative estimates. Equally if a long-run job is printed for call-off at regular intervals, the cost of warehousing it must be considered: space costs money. If the length of run is not known exactly, the best course is to ask for the minimum run likely to be needed and for a 'run on price' per thousand (more or less) for extending the run. If more than one paper could be used for the job the cost and samples of the alternatives should be obtained as part of the estimate.

USING THE BUDGET

Wide variations between different printers' estimates for the same job have long been a matter for discussion among print buyers. The reason for disparate estimates could be that, though all printers work in the same industry, they make their profits with equipment suited to particular classes of work. Rather than turn jobs down they may deliberately quote high prices for jobs outside their normal specialisations and leave it to other printers to undercut them if they can. The bigger and more closely specified a job is the smaller the variations of printers' costings for it are likely to be: no experienced print buyer would hawk a large, important and complex order around the marketplace looking for a marginal cost saving alone.

Some variation in estimates can be explained by simple logic. If the same specification is not given to all printers asked to quote, estimates will vary and will not be comparable. On occasion printers may temporarily have spare capacity which they want to fill and, to do so, cut their profit margins for a job which arrives at this time. Machine capacity is one of the things customers pay for. Idle machines not only fail to make money, they also cost money while they are idle, and few large plants are invariably working to full capacity.

It is legitimate for buyers to exploit this fact where possible by scheduling work so that it can be placed when spare capacity is available. An example is the printing of travel brochures. Orders in this largely specialised and profitable category of print are usually placed later in the UK than in other parts of Europe. It is therefore possible to predict when and where spare capacity is likely to be found and schedule a job accordingly, provided this can be done without risk: travel print runs to extremely late and tight schedules.

For some time, fluctuations in currency values have enabled very competitive pricing by overseas suppliers which, for large jobs, have encouraged buyers to look for cost savings by placing big print orders abroad. This was to be expected, but a cautionary note must be sounded if, on printing costs alone, a buyer is tempted to shop overseas for bargains. Apart from the risk of currency rates changing and wiping out the price benefit, freight costs and duties can at present add as much as fifteen to twenty per cent to printing costs for a large order. There are also longer-term risks. If the home market is deprived of orders by overseas competitors it cannot find the money it needs for re-equipment and continues at an accumulating disadvantage. At worst (as in the case of book printing in colour) UK capacity simply disappears and buyers are forced to look elsewhere for their print needs, even when there may no longer be any cost benefits to be gained from doing so. The commonest reason for final costs exceeding estimates lies, however, in-house, for instance in extensive alterations and revisions to the original specification, or at the proofing stage. Some precautions can be taken to keep a job to budget. Materials costs will be indistinct if the run length is not accurately calculated, or is altered after the order has been placed; materials can account for as much as sixty per cent of the total price of a job.

The buyer should be alert to the difference between fixed and variable costs: an 'agreement' on a quote is no guarantee that the work will be done at a stated price, though a written agreement on the terms of payment is worth preparing for jobs of any substantial size. The buyer should, I believe, resist any request for advance or 'staged' payments for all but long-run, long-term work.

Anyone who buys print enters into a contract with the supplier which is binding to both parties. If a dispute arises it will be harder to resolve if the buyer does not know what the contract implies. A contract can be written or verbal. In both cases the acceptance of the supplier's offer to do the work, if unconditional, has legal status. The Sale of Goods Act states that even if a price is not specified in a contract, or cannot be agreed between the parties to it, once the job is finished the customer must pay 'a reasonable price' for it.

Most printers have standard conditions of contract, which

sounds simple until it is discovered that the conditions may be either those recommended as 'a fair and reasonable way of doing business' taking into account 'the well-established practices and usages of the trade' or 'standard' only to the firm which makes them.

The buyer of expensive print should be aware of the kind of contract he is making. The British Printing Industries Federation has a *Standard Conditions of Contract* booklet which clearly sets out its recommendations and meets the regulations of the Office of Fair Trading, but a firm's membership of the BPIF does not necessarily imply its adherence to the Federation's recommended terms.

Bearing in mind that the *value* of print to the buyer can be higher than its cost, insurance and legal aspects are ignored at some peril. If a mail order catalogue or a price list is impeccably printed but delivered so long after an agreed date that the buyer suffers a loss of sales, the law can be invoked to compensate for that loss. There are also provisions in the Sale of Goods Act covering the ownership and care of original materials, such as paintings or photographs.

The buyer's obligations are also legally defined. They include ensuring the arrival of origination and the passing of proofs on schedule. The printer, by the way, is not liable for errors in the proofs, and author's corrections are normally charged for. The buyer is also responsible for supplying origination which will not entail more work and time to process than is provided for under the schedule for the job – for example if it is hard to read and transcribe ('poor copy' as printers call it) – though a legal definition of what constitutes 'poor copy' could be hard to establish.

Once a machine is running, the unit cost of each successive impression slowly diminishes as the run length increases. 'Split runs', where the same job has to go back a second or third time to the press, may therefore be more costly because they involve additional time needed to set up the machine for a new run and, for some jobs, more materials waste.

USING THE BUDGET

Recent technology using computers to pre-plan, proof, store and output data, which can then be read into the press controls and used to set such variables as inking and damping, can have positive cost and other benefits to buyers. Pre-press systems vary, but will include most of the necessary conditions for getting a job onto the press and running it under programmed control. This allows the buyer to avoid increased costs for repeat runs due to the need for re-setting the press for reruns. Publishers in particular have obtained solid benefits from computerised pre-press systems by specifying minimum runs and not having to pay a premium for reprinting. This way they do not need to warehouse printed sheets for call off and, maybe, lose money if the book doesn't sell well and the stored, paid-for print is not needed. On the other hand, a book which sells well can be back on the press for a reprint very quickly, using the computer-stored data which was created for its original printing, and the publisher is, for the first time, in a position to specify minimum initial runs and obtain reprints with calculable savings.

5
Selecting suppliers

It can be no easy matter finding the right printer for the job in hand. When a printer is chosen the buyer must give careful thought to the way a job is originated, progressed and monitored, since origination and production methods can have a good deal of influence on how efficiently and economically it is done. Indeed, there could be alternatives to having the job printed commercially. The much-publicised 'paperless office' actually requires an amazing amount of paper, much of it printed and some of which needs printing and circulating in various quantities internally and externally. It may also have computers, copiers, word processors and printing presses. It is, however, not planned and equipped to do everything for itself. The benefits of commercial printers are usually speed, cost and quality. Now that there are, for example, specialised business forms printers, the buyer can use them, provided he knows what to ask for, and is prepared to measure his specifications to what can be done on their equipment.

For some purposes, the buyer will use a 'panel' of suppliers (printers, materials suppliers and trade services) built up by commonsense and experience; but to take this course inflexibly, stolidly refusing to see representatives from unfamiliar suppliers and ignoring technological change, is risky.

It is time-consuming to see every rep who calls, but the remedy is simple: know enough about the organisation's present and prospective print needs to operate a fast-selection system. Companies and their sales people differ from one another, but those who just walk in or telephone asking for orders can be dismissed. Any company worth its salt will have researched what it believes the buyer is looking for and prepared concrete, useful suggestions and reasons why it should be considered as a supplier. Nowadays these might well include the availability of new equipment which could improve on existing services and be able to demonstrate a solid benefit to the buyer.

Though price is important, the buyer who puts it first and last is forgetting that he is purchasing more than print on paper: he is also getting – or, if he is not, *ought* to be getting – a high level of co-operation and advice from his suppliers. The print buyer who shops on price alone will, of course, end up with the cheapest suppliers, but not necessarily with those who do the best work; and if a job is mistimed or goes awry in some other way the known and trusted supplier is more likely to respond to the situation and to do what he can to retrieve it. In the present competitive climate in printing, inexperienced buyers can be deceived into thinking that spare capacity and 'marginal costing' (the polite term for price cutting) are the order of the day. So they might be in some categories for a while . . . but not for long. Companies who need to accept unrealistic profit margins are not among the ones most likely to stay trading.

Do not use the 'blunderbuss' technique of getting estimates. Buyers who send specifications to every potential supplier make work for themselves since, when they arrive, all the estimates have to be analysed and compared. There will always seem to be a cheaper printer somewhere, but the cost saving could outweigh the expenditure of finding him. An experienced print buyer I know said: 'When a hungry printer agrees to lower his price, another hungry printer is told as much and has then to drop his price below his competitor. When the job is eventually placed the printer who has submitted to this treatment will do everything he can to avoid losing money, which means he may cut corners. The project which starts out unhappily will get worse as it proceeds.'

How records are kept will depend on the range and nature of the work. The main requirements for keeping adequate records of printers are few. They could include (a) known printers who have already done satisfactory work for the organisation, with a list of their plant, a note of any specialisations and, perhaps, one specimen of each job done attached to the paid invoice for doing it; (b) printers who have done occasional work, with the same information; (c) printers who may not have done work for the buyer, but who have approached him with interesting or useful proposals; (d) an up-to-date list of trade services. Simple though these records are they can be neglected or deficient if not taken seriously. If neglected they get out of date, particularly in the area of plant listings. Printers and trade houses – even good ones – vary enormously in the amount and detail of information they provide about their plant and its capabilities. I know of printers who have invested in expensive new production equipment and failed to inform their existing customers, let alone potential ones, that it is there.

Do not neglect 'unusual' printing processes or new applications. While offset lithography now dominates commercial printing, other processes such as die-stamping, thermography, screen process and flexography, have shared in technological developments which make them valuable in ways which might not always be identified by buyers who have written them off in the past as being unsuitable for their particular categories of work. Flexography used to be considered primarily a process for cheaply-produced packaging, but it is now being used for the colour printing of newspapers. Screen process, in particular, can demonstrate some remarkable new tricks. Catching up on unfamiliar processes and their development is essential homework for the conscientious print buyer.

For direct supplies of materials such as envelopes or

wrappers, the printer will probably be able to make suggestions, and his suggestions should be considered: they will be based on a knowledge of the job and the risk of such obvious (but common) errors as ordering envelopes of the wrong size is avoided. Merchanting services are valuable for making cost and quality comparisons.

There are now services available for addressing, enclosing and despatching printed material. Whether or not they are used depends on the size of the job. If they are computer-based (ie, use the customer's computerised data for processing address lists) it is essential to ensure that the data supplied is correct and up-to-date *before* the order is placed; their equipment cannot be held responsible for errors of input.

6
Working with designers

A print buyer spends a good deal of his time working with other people who have separate parts to play in the origination and production of print. If he cannot do so effectively and harmoniously he is making his own work more arduous than it need be. In some respects it is easier to work with designers, copywriters, typographers, artists, photographers and other specialists within the buyer's own organisation than outside it: they should be more familiar with the organisation's print needs, and with the procedures in which jobs are progressed. Outside collaborators could need fuller briefing and more 'hand holding'.

Both sources will contribute better and more useful work if they are told as much as they need to know about the job as a whole. Designers especially are often accused of working in ivory towers, divorced from the commercial realities of the job. If they are, they should be working for art galleries; but the print buyer can unwittingly encourage people to retreat into ivory towers by not acquainting them with what's going on outside. For example, if a typographer recommended a typeface which a letterpress printer did not have 'in case', the printer once had the alternative of buying it in or persuading the customer that another typeface was 'just as good', usually the latter. Phototypesetting has 'liberated' typographers and designers from many of the constraints of metal type; so much so that there is already a generation of typographic designers who feel that it is unnecessry to know what these constraints were. They are the ones who are most likely to experiment. It is necessary for the print buyer to act as an intermediary between a creative team and the production process when the latter is changed in ways which affect any member of that team. Concepts which cannot easily be translated into practice might sound exciting, new, original and a lot of other things: they are not always economical or, sometimes, even feasible. In accepting the limitations of process and production a typographer, photographer or designer is simply working to some purpose rather than in the dark, and ought not to complain or invoke 'artistic licence'.

Apart from information about what is planned and how it will be carried out, a designer will benefit from direct knowledge and experience of the ways in which the equipment translates his work into print operates, and what it can and cannot do. He might know enough about colour scanners not to present artwork mounted on heavy board for colour separations made electronically; but does he know what the scanners can do which process cameras can't, and how they can be used creatively? If not, show him a scanner and let him talk to the operator. Does he know much about printing papers and the ways they behave on presses? If not let him talk to the paper buyer, the printer, the merchant or anybody else who can enlighten him. Does he know that, under some circumstances, 'proofs' are obtainable from new systems which present them for modification electronically, provided he is present at the terminal? If not get him to read chapter 9 on colour proofing.

Design services offered by printers have certain built-in safeguards: since the typographer or designer is employed by the printer he knows a good deal about the equipment which will be used in production, and about capacity and scheduling too. He will therefore work more closely within the plant's capabilities than an outside designer and is less likely to create problems on the production side. This does not mean that he will necessarily produce better work than an outside designer – only that he will be less liable to 'rock the boat'. Suspect printers who offer 'free' design services. If such services are worth having they are worth paying for, and design costs should be separated from production costs so that the buyer can see what he is paying for. Hidden costs are always a worry to print buyers.

Try to avoid asking artists for 'roughs' before a job has reached a stage where the roughs have some relationship to the way the rest of the job is planned. When working in an advertising agency I saw more time and money spent on roughs and expensive photographs than on finished artwork! It was because some people cannot imagine *any* solution unless they can see it before them, and are given a choice of solutions, too. They paid for it. So will you.

7

In-plant printing

In-plant (or, as it is sometimes called 'in-house') printing has grown in size, scope and pace over the last decade, and recent technology has established it even more firmly as an internal service. Where it is available it should take its place within the overall planning and implementation of an organisation's print needs and, therefore, comes within the scope of, if not under the control of, anyone whose task it is to manage those needs.

In-plant printrooms vary enormously in size and in the kind of work they undertake for the organisations which support them, but some idea of their importance can be indicated by the one large local authority which, at the time of writing, is managed as a single unit for both internally-produced print and that bought outside. The department processes some sixty-five million pages a year. These emanate from a typing pool which processes around seven hundred thousand A4 size typescripts a year of which thirty-seven thousand are produced on word processors. The word processors have a throughput of some seven thousand pages a week, mainly documents which do not require conventional typesetting and printing. The average run handled by the internal printroom is low in commercial printing teams – from three hundred to five thousand.

Most local authorities, government departments and other public bodies, and many commerical and industrial companies, have in-plant printing facilities, but it would be impossible to say with any accuracy what is a 'typical' in-plant set-up. These units have to be competitive with commercial suppliers for the work they do, or demonstrate some other clear reason for being maintained. The print buyer's task remains that of getting the required print produced cost-effectively from whatever source it comes.

With good management an in-plant service can play an important part in the print planning of an organisation. To this must now be added the growing influence of the 'electronic office', with its word processors, computers and other systems which can, and very often do, become involved with material which will be forwarded for printing. If a large volume of text origination, or 'input', is continually being processed it can be an expensive part of a printed job, while the cost of 'output', in terms of printed impressions, is relatively low.

In-plant printing started with the availability of small offset-litho presses and ancillary equipment which became widely available in the 1960s and grew, with office copiers, to replace conventional duplicating machines. The in-plant proposition was then straightforward: regular supplies of simple print, such as internal stationery and forms, could be produced more conveniently and, possibly, more cheaply in-house than outside. The decline, and virtual disappearance, of local jobbing printers stimulated in-plant developments. The benefits looked easy to calculate. They included lower print bills, though some in-plant departments then, as now, were poorly costed. Many were set up and run with the haziest knowledge of, and control over, overheads, such as the cost of the space they occupied and the materials they consumed. These were not so much managed as allowed to grow until somebody noticed what they were costing for what they supplied. There were, however, enough of them doing useful work to encourage manufacturers to cater for their special requirements.

Automatic platemaking, small bindery systems and improved presses came along. The commercial printing industry looked on with official disfavour, but could do little about it. For many print needs the in-plant units could not be expected to cope. Few, except some specialised ones which virtually resembled commercial printing plants, could handle typesetting or colour work. Most knew their limitations and kept to them.

What has now given a new lease of life, and new directions, to in-plant printing, and materially widened its scope, is the latest technology for the origination of text and graphics. There are ways for converting the output of word processors into formatted typesetting. The increase in the number of trade services which will handle such origination and other requirements, such as binding and finishing, is a further stimulus to the in-plant unit to become more adventurous in the kind of work it undertakes. To this can now be added the enhanced speed and capabilities (including colour) of copying using the latest machines on the market; but before it all starts to look too glamorous some broad limits might be set for what can sensibly be projected for the future of the in-plant unit.

Basically these are defined by what the printing industry is equipped to do more quickly, more satisfactorily and more economically than an in-plant printroom. The latter is unquestionably a cost centre in capital equipment and operating overheads, and one which is not easy to control. If the printroom is seen as an 'office service' on a par with the coffee machines or window cleaning it will be an extremely wasteful one.

The print buyer should apply the same yardstick to a printroom as he uses to select external printers; quality, cost, service and delivery, in whichever order these are important to the job in hand. The worst thing that can happen is to allow the in-plant unit to 'have a go': the buyer must know its capacity and capabilities and not waste time and materials passing work to it which it cannot handle efficiently and economically.

Technological changes in text origination

All text required for printing has to be originated in some way. Once there was only one way: by writing or typing it so that it could be transcribed by a compositor for conversion by typesetting machines into printing type. This applied even after the development, some twenty-five years ago, of phototypesetting. The only difference was in the product of the machine, which was type images on photographic film or paper instead of metal type. Computerised phototypesetting was the next step, which enabled type to be composed under the control of computer programs which inserted many of the commands required to set the type in the size, style and format required. They also, with varying success, made the word-breaks at the end of lines which, in accordance with certain rules (and many exceptions to those rules) are used by compositors when setting justified text.

In these, and subsequent, developments in typesetting the need for transcription remained: there had to be 'copy' and it had to be copied. The exception was where an original typescript could be used direct to make a printing plate photographically. This came to be called 'strike on' composition and is still used. It is, of course, not capable of creating a page of printer's type, but simply of reproducing, as do copying machines, what is already on the paper. Where the text has to be reproduced as printer's type strike on composition is impracticable.

Throughout the development of phototypsetting a good deal of attention was paid to the problem of 'capturing the keystroke'; the jargon is (for once!) self-explanatory: if text can be 'captured' for phototypesetter input as it is written then time and expense are saved by not having to rekeyboard it. Technologically the problem was harder to tackle than it first appeared. In addition to the text itself its transcription calls for informed decisions on how it will look in print and the skill to apply them at the keyboarding stage. Typefaces, and their sizes and variations, (boldface, italic, etc) must be catered for in computerised typesetting by 'commands' (usually called function codes) which instruct the computer to do what is wanted. The rules for wordbreaks, fount changes and other typographical subtleties must also be implemented by the typesetting program within a determined format. A problem existed in making corrections and alterations as a result of the spatial (format) changes which these generate. Additionally the quality of the output of phototypesetters, until quite recently, often left a good deal to be desired compared with metal typesetting.

Data processing technology lagged behind the typographical demands made on it for many years. It was often difficult, or impossible, to demonstrate any immediate saving in cost or time in computer phototypesetting for volume text composition since the fundamental requirement remained the transcription of the copy, the insertion of the typography and doing all, or most, of the things done in conventional typesetting; the obvious value accrued only at the updating stage for text revision before reprinting.

Developments in computerised typesetting followed one another increasingly quickly throughout the 70s and 80s. It was eventually found that, provided input existed, computer-held data could, with some technological ingenuity, be used as *typesetter* input.

With the advent of 'mini' and 'micro' computers, and the consequent size and cost reductions which they made possible in systems design, the whole typesetting area took new directions and led to further computer-related applications in the typesetting of text. Some of these relate to the production of full pages incorporating both text and graphics.

What it comes down to is this: any text which is capable of being stored in computer-accessible form can, with the appropriate equipment, be output as typesetting *without* intermediate keyboarding. That means any text produced by computers or word processors (which store text on disc), even small personal computers, is capable of being used as input to phototypesetters which output finished, formatted setting. The implications are many, not the least of which is that any appropriate keyboard in any location is theoretically a 'typesetting' keyboard; but this is not always the case in practice. Keyboarded input can be originated anywhere and conveyed physically, or transmitted by communications links, including telephone lines or even radio and satellite links, to the production end. This is itself a dramatic change from the time when all copy had to be routed through a succession of production stages to convert it from an original text into a typeset page.

It has, however, created a new set of conditions for doing so successfully. Modern phototypesetters using only their own typesetting programs are not capable of carrying out everything a trained compositor could do. All computers in phototypesetters have to be 'told what to do next'; they are versatile, but they are not intelligent, nor can they make aesthetic judgements. The typography and format of the typeset output must either be inserted into the original keyboarded input from such machines as word processors, or incorporated later as an addition, in the form of coded commands, into the keyboarded text.

At present this is tackled in two possible ways: passing the stored text to the typesetting terminal, where it will be read on a VDU and the necessary phototypesetter codes implanted by a skilled operator in accordance with the programmed requirements of the phototypesetter and the typographical design of the job, or directing the keyboarded input to an interface which also has a small computer which will automatically insert the function codes for the phototypesetting run. Both methods tackle the same problem which, in essence, is to relate the operating code structures of phototypesetters, word processors and other kinds of computer-generated text to 'raw' text input.

In practice this is not easy. Text for handling in this way

must be accurate – otherwise what is input to the typesetter will be incorrect. Somebody has to decide *at the text origination stage* exactly what is wanted in type and also ensure that a full typographical markup is available at the 'typographical editing' station; the value of the whole concept depends on not having to rekeyboard *any* of the text. This demands an extremely high standard of original of the text supplied to the typesetting terminal.

Word processors can revise, correct, update and store on disc any keyboarded input, and print it out as many times as required. This, for practical purposes, is a 'proof' of what will eventually go forward for typesetting. It does not, of course, look like type on the page because it has not yet gone through the typesetting process. Unfortunately many people cannot visualise typesetting unless they can see it: that can include designers, and some print buyers! But unless the buyer is prepared to accept the word processor printout as a 'proof' – at least for the purposes of ensuring that it doesn't contain mistakes – the benefits of 'direct entry' systems are negligible.

Once input has been typeset and made up, all corrections, reformatting, repagination, etc, will have to be re-run through the machine for corrections and alterations to proofs. It might not appear expensive to use a phototypesetter in this way to provide a 'hard' proof, though it could become so in terms of time and materials for large amounts of typesetting compared with the 'soft' (word processor) proof which is available before input. The technology passes back to the customer for primary text origination though it must be borne in mind that a proportion of text has to be keyboarded in any event.

In this, as in other areas where the customer assumes a measure of control over, and responsibility for, what appears on paper, the benefits looked for are likely to be speed and cost savings; but neither comes automatically. First, is the equipment installed and in use at the buyer's end? If so, are there sufficient resources and skills to hand in order to apply it to typesetting tasks efficiently? Word processors are not designed as typesetting terminals and it quite surprised their manufacturers when such applications were mooted. Their primary functions – handling documentation and correspondence – could well come first, and the system might not be worth using for such added functions. In many cases – especially where there is an in-plant department – it could be more sensible to consider installing a low-cost phototypesetter now that these machines are available. Either course would call for a new approach to in-house text origination.

The most important question to be answered when considering various ways of using direct input is when and where coding structures are to be introduced. It would be too much to expect an author or a word processor operator to insert all the function codes which would guarantee the output of all typesetters. The writer is not normally also a typographer. One course would be a system of marking up word processed copy to indicate what the designer wants done at this stage. This would be no different from the typographical markup which typographers supply.

Standards for 'text preparation and interchange' are being drafted but have not at the time of writing been published by ISO. And, of course, a whole raft of standards authorities are beavering away and will no doubt come up with the usual sets of differing recommendations.

What, in the interim, is most needed in the direct input field is a simple, comprehensible, and fairly comprehensive, set of marks which can be used on copy to indicate what has to be done to arrive at the desired typographical output. Remember that the 'black boxes' – the interfaces which change word processor output into phototypesetter input – cannot make *decisions* about what the phototypesetter does with the input.

A useful guide has been prepared and published by the British Printing Industries Federation. It is a relatively short list of codes which forms a reasonably flexible mark-up 'language' for the most widely-used kinds of formatting and typesetting. It is called ASPIC, a mnemonic for 'authors' symbolic pre-press interfacing codes'. As the Federation observes, 'A comprehensive scheme covering all the nuances of typesetting will take much longer to develop.' ASPIC comprises a number of basic codes or 'flags' which can be keyed in with text and which show a compositor or reading device any change in typeface, layout, etc, specified by the author or designer at any specific point in the text. (Remember that 'reading devices', if they are computerised, can be programmed to recognise virtually any formal set of instructions which are *consistently* used for the same purposes). Encoded text can be input to a word processor or micro computer and printed out as an adequate typographic proof before a final decision has been made about typographic style: if changes are made the codes are changed before input to the phototypesetter.

Essential ASPIC is a short list of codes which caters for most routine jobs comprising text and paragraphs, headings and, perhaps, a few typestyle changes such as emboldening or italicising. There is also Language ASPIC, Tabular ASPIC and Mathematical ASPIC which cater in additional ways for these requirements. Practically all the codes comprise two characters or a character and a figure in brackets, such as [hl]. Each bracketed pair instructs or initiates a specific operation, such as centering, indenting, font changes, space adjustments, and so on.

ASPIC is a brave but temporary attempt to get this technology off the ground without waiting for longer, more ponderous and probably less comprehensible standards where every tilde, umlaut, and cedilla becomes part of a vast map through which users try to find their way to what they want – which might be no more than how to indicate a simple paragraph indent!

Trade houses again come into the picture. Many such services have been started to meet the demand for interfacing word processors with typesetters. The use of word processor bureaux in tandem with trade phototypesetters is now commonplace and, as with all services which can be bought in by the customer, benefits are assessed mainly on speed, cost and convenience.

Direct input systems are 'front end' systems; that is, they do not carry out the setting, they *initiate* it. If text throughput is high the buyer could well find himself in a queue for typesetting capacity. It sounds impressive – as, indeed, it is – to learn that text can be sent 'over the phone' (using a telephone modem) to a typesetter, but the job isn't done until that text gets to the machine and, fast though they are, phototypsetters can only handle one job at a time.

8
Computerised graphics

For present purposes 'graphics' – a term widely and loosely used – will be taken to mean all visual elements except continuous text. It includes illustrations in line or tone, headlines, logos, photographs and decorative devices. It is usual, but less exact, to call graphics 'artwork', which some are, though not all.

For graphics, the customer conventionally supplied origination (film, colour transparencies and made-up pages) in a highly finished form, ready for reproduction, just as text was once invariably supplied as 'copy'. Graphics may include a certain amount of lettering, and small amounts of text when this is an element of graphic design; headlines, for example, are frequently originated by graphic designers using drawn or transfer lettering.

For any job with a graphics content, the print buyer will connect people who originate graphics with those who reproduce them for printing.

Where the organisation has its own studio and art direction, a job will reach the print buyer in an advanced stage of readiness for reproduction. Advertising agencies work in this way, as do the advertising departments of large firms. The print buyer controls the progress of the job and ensures that it is carried out as it has been planned.

Techniques in graphic design are many and varied, and are a study in themselves, not only for designers but also for print buyers who are concerned with how material is presented for reproduction. Many of these methods and techniques are described in *The Graphic Designer's Production Handbook* by Norman Sanders which I edited for a new British edition (David & Charles, 1984).

It is a recipe for disaster for a print buyer to assume that a designer, an artist, a photographer, or anyone who contributes a visual dimension to an item of print can be left in ignorance of how it is intended to produce it.

In some respects graphics can be more straightforward than text from the print buyer's standpoint. The equipment and procedures used for reproduction in offset litho printing are precise and flexible. Provided the buyer's specification is informed and complete, he will reach the repro stage by shorter steps and with less trouble than, in many instances, he could with text processing: graphics are visible in their *originated* forms as photographs, artwork, layouts and so forth. Reproduction standards can, with practice, be established by comparisons and measurements. Some of these will be examined in the section on colour printing and proofing (see chapter 9).

Using conventional origination – the assembly of graphic elements within a format for platemaking – it might be necessary only to ensure that the origination is supplied to the trade house or the printer in correct sizes and with a foreknowledge of what systems of reproduction will be used.

The most radical departure from the established conventions of graphic origination is in the equipment now available to do a number of the things needed to prepare it for printing. These systems are mainly computer-based; all demand radical changes in conventional graphic design procedures. All can be used by the customer in his own studio or by a printer or trade house and all fall broadly into two categories: first, the 'previewing', editing and formatting of graphics and/or layouts and second, storing the design as electronic data in a computer which can output it in usable form for repro. Computerised front-end graphics terminals permit the graphics to be brought to an advanced, or even complete, state of readiness for photomechanical reproduction before printing plates are made.

A fairly straightforward example is the business forms makeup terminal, which serves to illustrate the principle of other, more versatile systems. Forms makeup terminals allow a form to be composed on a screen, using a combination of elements normally required in a wide range of business stationery – lines, grids, boxes, tints, text, etc – which are accessed from computer store. The VDU will illustrate the designer's instructions and change or modify them on command. Commands can be given from a keyboard, or from a hand-held graphics input device which is moved over a design and used to pick up the electronic information and transfer it to the computer. It can be viewed as it is assembled. This sort of design terminal allows changes to be made, and seen, for as long as the operator wishes to make them. It is capable of producing an 'interim proof' of what has been composed and it stores data in a retrievable form. It will output data as film or paper and can be used to make a printing plate.

A significant development in both text and graphics terminals has been the way in which data are stored. This is described by the unlovely word 'digitisation', which means simply that the computer stores graphic data as figures, not as images: if you depress the key marked 'X' on a computer's keyboard the computer does not place into memory the letter X, but its coded equivalent in computer language, ie as 'digits'. Any shape, be it a line, a character in an alphabet, a logo or other image can be stored thus and accessed and output without being held in computer store as an image: it is *assembled* from digitally composed data in computer memory under programmed control.

At one time all phototypesetters incorporated discs or matrices on which were held, visibly, all the characters needed for the setting. The computer contained none of this graphic information within its typesetting program. Digital phototypesetters and graphics terminals do not require this physical access to character forms: these are part of a computer program which is written for and 'dedicated' to specific graphic uses.

This brief excursion into computer behaviour is worth taking for it leads to an understanding of what a lot of today's computerised origination equipment does and where it is leading. It is no easy matter to map images as precise as typefaces in a digital computer and to position them precisely. Digitisation of type and images is the most recent and, in many ways, the most important development in graphic technology, and points the way to extremely powerful and versatile systems.

The example of the business forms composition terminal can be fairly easily extended to what has been called a 'graphics workstation'. Typically this terminal might include a 'forms builder', a graphics builder for the creation and 'editing' (modification) of graphic images, which scans an existing image and displays it at the terminal for editing, and a text generation and editing terminal. The operations possible at the terminal might include access to a stored 'library', graphic elements, enlargement, reduction, 'cropping', resizing and layout of complete pages. The operator can try out variations, experiments and ideas, and preview them, or use them as interim output ('proofs') until he (or a client, print buyer, studio manager or anyone else concerned with the appearance and content of the printed product) is satisfied that what is shown is what is wanted.

An extension of these previewing and editing techniques is in the area of colour photographs or artwork for printed reproduction. A colour image on a domestic TV screen is made up of lines and the colour image reproduced would be far inferior in colour fidelity and sharpness to that needed for graphic reproduction. If the data includes colour pictures there are new requirements: the elements which combine to produce colour combinations must also be capable of digitisation as 'cells' (or, as they are called 'pixels'[1]). This is possible but the colour data must also be 'read into' the computer. It cannot, as can type, be input from a keyboard, or, as with forms composition terminals, be held as library graphics in computer memory.

Colour previewing and makeup therefore need the input of a scanner which 'collects' the data from the original and transmits it to the computer where it is digitally stored and available for output to the graphics terminal, allowing the modification and correction of colours and colour balances shown on the high resolution VDU. Thereafter the modified, scaled picture is returned to computer store for output to the colour scanner which will make the separations in accordance with the stored instructions and under the control of the program. This is a considerable simplification of what actually happens but is sufficient to show what the equipment is designed to do.

All these items of equipment can be used separately, or in combination. Where they are combined to supply several kinds of input/output at the graphics terminal we have a full page makeup system, which allows a whole range of graphic elements to be accessed, separately or together, and combined to assemble (lay out) complete pages. These pages, returned to a scanning station, will control the scanner to output film separations which incorporate not only the colour pictures, but also any other graphic elements in the page – text, rules, tints, headlines, etc.

All this raises new opportunities. It could replace the conventional techniques, materials and procedures of the designer's studio and the conventional colour repro workshop,

---

[1] Pixels are part of 'bit mapping' which is used in several types of systems where graphic data has to be stored. It is also used in colour scanning (see p71) in connection with raster image processors (see p111). These 'picture elements' are made up of dots, and measured by the number of dots to an inch. Thus a pixel measure of 300 × 300 would mean that the screen could reproduce 300 dots, laterally and vertically, to one inch. Each pixel is individually addressable, so that it is possible to arrange the matrix to reproduce fine lines, type, halftones and even handwriting.

but it replaces them at considerable cost in equipment, and its users will need to understand it and adapt to the methods dictated by the equipment. Computer-aided design is a designer's tool and should be used as such.

The graphic designer should be involved with it wherever it is being used. Whether he will want to is another matter. There is no doubt that, where computer-assisted origination systems are employed, they are not substitutes for designers' skills: computers are not 'creative'. Designers, photographers and artists (and one could include authors) are trained and work in particular ways. Maybe a new generation will welcome the ways technology helps in carrying out certain kinds of creative work, but not as electronic substitutes for human creativity. The systems are, in fact, aimed primarily at exploiting the computer's power to save time and money and to reduce manning in labour-intensive pre-press areas.

Few computer assisted design terminals find their way into customers' premises. At the time of writing this no more than two per cent of graphic designers in Europe are using computer graphics workstations. One authority, Dr Lionel Wardle,[1] puts it thus:

They [the potential buyers and users] would rather forego any advantage than risk buying something that might be cheaper later. It's understandable but not very clever. People invest in computer graphics because they believe it will improve productivity, reduce costs, create new opportunities or put them ahead of the competition. Compared with the earning potential of a system, changes in prices may well be seen as marginal or even irrelevant.

A summary of the characteristics of a graphics workstation could be useful. It could comprise (a) an input device to convey information to a computer, which could be a light pen, a keyboard, a 'touch' screen or a camera, (b) a computer, (c) a display screen to present images, including text, (d) a data

---

[1]*Graphics World*, January/February 1984

storage unit, such as a disc store, (e) an output device to produce 'hard copy', such as a printer with graphics capability or a camera/recorder, and (f) programs which organise the operating system and interface with any devices connected to it.

Full-page makeup

The four basic questions concerning a page makeup terminal (PMT) concern the ways in which this development is likely to require informed decisions by those who use them. They are: 1 How much interactivity should the system have? 2 What degree of typographical detail should be displayed? 3 How and by whom will the system be used? 4 Should the system be totally integrated or made up of subsystem components?

The answers to these questions will depend on the kind of work envisaged for a PMT. If a PMT is not part of the in-house resources of an organisation, such equipment still has a direct bearing on the ways in which print is progressed by the print buyer. He may not have the final decision on whether such equipment is put to use, or where it is located, but in either case it will affect his working systems and call for closer technical liaison at origination stages than might be necessary under conventional systems.

The four questions involve both administrative and technical factors. The first, concerning the degree of interactivity, relates to the possibilities which now exist for using a PMT for all pre-press work, even including platemaking, and envisages the origination sector as, effectively, the major, if not the sole, input to the printer. The latter's job would, in such an event, become simply one of factory production – the use of his presses to produce the required number of printed items.

If, then, the PMT is to be used by the design and layout department in-house it must be remembered that such work will go much further than the normal design and preparation of artwork etc for subsequent processing: it will create input which is ready for production and which leads directly to platemaking and production.

The second question – how much typographical detail is needed at the PMT – relates to the first. Some systems present only a representation of text, and do not include typographical refinements such as precise spacing, exact representations of typefaces and other typographical details. The 'window' of the VDU does not allow them a complete view of what is inside the system.

How and by whom should the system be used? In some situations there is little choice in using both origination and production facilities in the same place. A newspaper is an example. It cannot be 'originated' (written, sub-edited and made up with headlines, text and illustrations) and allowed to take a leisurely course to production. There has to be a short, direct route from origination to completed production.

Some of today's new equipment presents the possibility of at least a comparable degree of direct input. The presses, finishing equipment and typesetting equipment remain at the production end – the printer – and it is occasionally forgotten that, dramatic and impressive though much of the page makeup technology is, the actual production equipment is basically similar to that which has been used for some time.

Few want to buy expensive, advanced equipment simply because it is available, or to install it without clear evidence of its contribution to greater economy and efficiency, and the speed and accuracy with which it will carry out specific tasks. Thus PMTs may well be used in-house by organisations which are continually involved with the creation, origination and processing of type and images; advertising agencies, for example.

In general, it can be assumed that when systems are used for origination in-house the reasons for doing so will be clear and show a return on the capital and operational resources invested in them. The situation is complicated by the fact that, when we talk about 'systems', we do not mean single pieces of equipment which do specific jobs in specific ways, but a range of equipment. In brief, you can cut the technological cake in many different ways. This is what is meant by the subsystem approach. Modular design has helped many systems to be built up in various ways so that, for example, a complete typesetting system is not needed to take advantage of the new input methods described on pages 45–51, and a completely independent PMT is not required to go some way along the road to direct input of text and graphics.

9

Colour

So much of what we see in print and elsewhere is in colour that it is easy to forget what a luxury colour printing once was, and how it was conserved for particular subjects. We accept colour on TV, in newspapers and magazines, in books and in many other places where, a few decades ago, it would have been unexpected. It will do no harm to restate some basics about colour printing: they still influence colour reproduction, however varied may be the means of reproducing it.

The three properties which most directly affect colour fidelity and consistency over a print run are the quality of the originals (photographs, artwork, etc), the materials used (inks and substrates, mainly paper and board) and the printing process itself.

Colour on a surface has two visible properties, saturation and reflectance. The first is determined by the amount of colour laid down and the second by the extent to which it is absorbed or reflected before reaching the eyes. For simplicity let's take an example: a single red line on a white surface. First the medium used to make the line and the surface itself will influence colour saturation. A dense ink deposits a greater 'weight' of colour than a thinner, more diluted, one.

Considering reflectance alone we are seeing the pigment (colour) in terms of the amount of it which reaches our eyes. A white wall in daylight absorbs few colour frequencies and reflects many back; a dark wall absorbs more. These properties are evident in the differences between papers: a white, coated paper has high reflectance; a darker, creamier or uncoated paper has relatively low reflectance, and a black one hardly any at all.

So we have our 'artwork' – the red line – which is not just any red line but *that* one in *that* medium on *that* surface. If we want to print it with as many of the properties of the original as we can reasonably expect, it must be colour-separated, photographically or electronically. This would be necessary if there were more than one colour on the paper but remember that the paper also affects colour and must be considered. We have so far transferred some, but not all, of the properties of the original to photographic film. This uses dyes which are a different medium from the ink, paint or whatever was used to create the original. When the colour separations are combined they make up a 'picture' of the original, the resemblance to which will depend on the skills of the repro camera or colour scanner operator: both can change the colour radically, on purpose or accidentally.

To make a set of printing plates (one plate for each colour to be printed) the film will normally be re-photographed (maybe more than once if excessive enlargement or reduction is needed) and converted into a developed image on the plate(s). Perhaps, for our single red line, only one plate would be required if a high degree of fidelity were not called for. For 'full colour' at least four plates will be made; so further conversion is needed. A proof now taken on a proofing press would show how near we are likely to get on a printing press to the original artwork, provided the proof is printed with the same inks and on the same paper which will be used for the press run. It cannot, however, show us *exactly* what the production machine will print, impression after impression, during the run because there we are dealing with a new set of variables: printing inks, paper and press characteristics which, in offset litho, include an extremely fine balance of inking and damping on the plate. The amount of ink which is laid by the press onto the paper affects colour-intensity; the paper influences colour reflectance. As long as a precise balance is maintained the impressions from the press should not vary in appearance. Over a long run they could well do so and press setting would need adjustment to bring the colour balance back within the determined limits for colour consistency. Add to these variables the screening which is needed to break the image into dots and we are, potentially, very far indeed from the appearance of the original. Yet such are the techniques and skills of modern colour printing that it can produce a very high level of acceptance on the part of the viewer that what is seen in print on paper is an accurate representation of what was there originally.

The first consideration in colour reproduction is the quality of the original. A printer could once say with complete honesty that he could not be expected to reproduce on paper anything which wasn't already present in an original. This is no longer so. Colour scanners (see pages 71–4) can, indeed often do, make corrections which process cameras could not, including increasing the 'sharpness' (definition) of an unsharp original and redistributing colour intensity and balance. These are not rescue operations on poor originals (or should not be) but effects which have been made possible by the equipment itself. The quality of originals is nevertheless still a high priority in attaining good quality colour.

Colour is normally viewed subjectively: there is no way of being certain that what I call red is exactly the same as what you call red, even if we are looking at the same pillar box. The best we can do, if we disagree about it, is to measure and compare colours to standards, and it is essential that the same things are measured by the same standards. Not so long ago no colour viewing standard could be applied, but not because a standard did not exist; it did, but it could not be used because there were no standardised light sources which conformed to it. If it cannot be agreed what colours are present in an original it is unlikely, even with the most meticulous measurements, that agreement will be reached on how accurately they have been reproduced, and there will be requests (familiar to many repro houses and printers) for 'a little more blue in the sky' or 'less red in the flesh-tones'.

British Standards, European Standards and American Standards exist for the colour viewing of transparencies and reflection copy; it matters less whether they are identical than whether they are practicable and are applied consistently. If transparencies are chosen by holding them up to the daylight and squinting at them no standard can establish exactly whether what is printed is what was seen in the first place. Again, if a pressman passes a sheet in a viewing booth under standard conditions for the viewing of reflection copy, it should in all reasonableness be passed by the customer under exactly the same conditions. A colour transparency will never look the same as a printed proof because it is viewed by transmitted light and not reflected light, and does not appear on paper. For such reasons I believe that print buyers, studio managers, designers and all who have a vested interest in colour control should start by having light boxes and viewing booths on their own premises.

The British Standard for the visual assessment of colour printing is BS 950, Part 1 *Artificial daylight for colour matching and colour appraisal* and Part 2 *Viewing conditions for the graphic arts industry*. There is also an international standard, ISO 3664 *Illumination conditions for viewing colour transparencies and their reproduction*. BS 950 sets out the ambient conditions as well as the illumination of flat sheets (eg, over benches and facing a matt grey wall) where a standard illuminant is used in the general lighting as well as in the specially designed light source above the print. Light boxes for viewing film transparencies may also conform to the above BS and ISO standards. One such box allows for a side-by-side viewing of colour positives. The originals are placed on mounts with the printed reproduction beside them. Both are subjected to the same light intensity and colour temperature from illuminants which conform to the British Standard. Remember that these are *viewing* standards, and not directly related to other standards for the *control* of colour reproduction on the proofing or production presses.

Densitometric measurements can, and should, be made, since they allow theoretically exact measurements of colour density to be made. Unfortunately, and with the most advanced of colour measurement technology, it is still possible to obtain two, three or more different readings from the same number of densitometers of differing makes.

A densitometer is an aid, not a substitute, for visual inspection and should be used to check what is seen in a colour proof. First the standard must exist, since the densitometer doesn't create a standard, but measures what is produced as a proof against the printed job. Proofs may be supplied by repro houses – often in various locations and not where the item is printed – and the densitometer provides a common standard *for that job* for them and for the printer.

For the densitometer to work it requires colour control bars across the entire width of the printed sheet, which are normally 6mm wide strips, one every process colour. Some control bars incorporate exposure wedges to assist in correct exposure during platemaking.

Densitometers have been tools of the trade for some time, but the modern hand-held densitometer can supply a much wider range of information than its predecessors. One instrument currently available will read the following data from the colour wedges which are normally printed on the margins of a colour proof, and it can also be calibrated with other instruments of various makes: density of solids within close ranges; surface coverage and any variation therein (such as might have been caused by dot gain or trapping, see pages 78 and 81). It will store two independent values for each colour channel. These instruments are battery operated and portable. Some may have additional features, such as a compensating circuit for wet and dry ink readings (the reading from a wet ink is not identical to the same ink when dry). Some densitometers will store colour data from readings already made and display the readings as numerical factors on a small screen. Used intelligently the densitometer is a most sensitive and useful instrument, and a great friend to the print buyer.

Another standby of the print buyer is the colour swatches which relate to the various colour-mixing systems supplied by inkmakers to printers using these systems. Print buyers are not always aware of the full uses to which the swatches can be put, and still tend to supply the printer with specimen colours cut from them which have no numerical or other precise identification, and have to be matched by trial and error. The systems are straightforward enough. They present a set of coded colour mixtures – five hundred or more – printed to an exact specification which relates to the inks used and the proportions needed to make a given colour. They may also show the differences in colour which can arise from printing them on coated and uncoated papers and with a variety of screen rulings. As visual matches they are approximate, but used in conjunction with a densitometer it is possible for both printer and buyer to match colours with some accuracy. The print buyer who keeps these swatches to hand had better do so in an organised way. Arguments about whether a printed colour is 'the same' as a specified colour are usually inconclusive. Colours can be affected by mechanical conditions on and off the press and by papers and screening when used with mechanical tints, so any attempt to match solid colours where tints are used is futile.

At best colour matching is an imprecise art, and this is realised by the printing industry itself. It now looks likely that work being carried out by printing research organisations will result in more exact ways of identifying variables in colour printing which can occur *on* the press and provide guides for controlling them within recommended tolerances so that, with a set of separations, it will be possible to operate machine controls to maintain consistency throughout proofing and printing stages and in the final products of the press. The standards set by BSI and its German equivalent DIN are only partially satisfactory. The mere existence of standards does not mean that they can be attained and maintained.

A more realistic approach is that of the printing industry itself, which has to deal with variables under production conditions. These attempts to define 'systems of best practice', as with all efforts towards standardisation, depend for their success on general agreement and wide acceptance. The best the standards-drafters have been able to do up to now is to hammer out a reasonably exact set of guidelines for a wide range of print produced under a variety of conditions to a broad spread of quality criteria. The customer can but wait and see what these 'systems of best practice' turn out to be and, in the meantime, continue using whatever controls are already available. The problem is not an easy one, since 'quality', like beauty, is in the eye of the beholder and, more to the point, quality standards vary depending on the job, its cost, and its functions. Nobody would suggest that a handbill for door-to-door distribution need be produced to the same quality criteria as a reproduction of a painting for sale in an art gallery. Many jobs do not call for top quality or facsimile reproduction, however that quality is measured or defined, and buyers would not welcome the additional costs involved in attaining it for work which is required cheaply to less critical standards. In the event it is probable that where standards are applied, it will be among the relatively few, large printers which specialise in long-run work to strict colour reproduction criteria, such as with mail order catalogues and some periodicals.

Proofs

With such problems in mind the print buyer might decide to remain on the sidelines and wait until the proofing stage where there is at least something tangible to see and compare. The word 'proof' is now a capacious one, and may include the electronic previewing systems described on pages 54–8. Where such systems are used practically all alterations and modifications called for can be made before ink touches paper which, broadly speaking, is what these systems are there for. In the section on colour scanning it will also be seen that the stages at which a 'proof' can be produced can be much earlier than they were when a proof could be obtained only from a printing plate.

More common than electronic proofing are photo-mechanical systems which allow proofs to be made and corrected to acceptable standards, taking into account some of the variables likely to be present in production and likely to influence colour fidelity. There are now several systems. The first to be developed, and now produced to European standards, is Du Pont's Cromalin proofing system, which illustrates what these systems are designed to do.

The reference source is a standard colour strip, or 'control strip', which has thirty-six squares which show all the standard process colours – cyan, yellow and magenta, plus black – in a series of patches giving dot percentages between twenty-five and seventy-five per cent, used by the printer to measure print contrast. Other patches show solids which indicate the density of the full tone and a range of tonal values obtained by various distributions of the solid over a white ground, with coarse and fine screen patches which allow dot gain (see page 81) to be checked.

The lower margin displays various combinations of the process colours from which can be seen the degree of ink receptivity of the colours (trapping) and the grey balance, plus highlight and fine-line patches which visually indicate exposure and allow a readout of colour balance.

Thus, with a strip measuring no more than 12.5cm × 2.5cm, a large amount of information is available and can be used, with a magnifier, to make comparisons on the proofed sheet. The colours on system-made proofs are not produced by inks but by toners, which are compounds deposited in

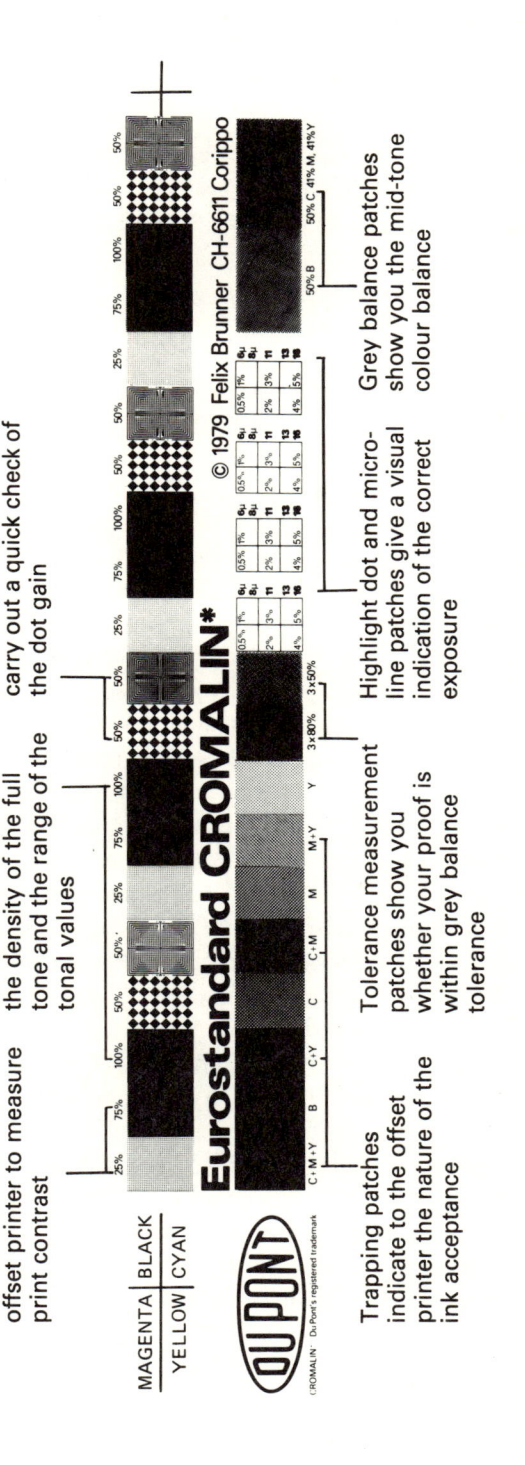

Du Pont's Cromalin photo-mechanical colour proofing system

measured, formulated quantities related to the conditions which will be present in production and to the inks themselves. The systems are not perfect, but have been developed to an advanced stage. They have a special value in long-run web-offset work where it is not feasible to press-proof a job to reliable colour standards.

Indeed the question of what is a 'reliable' proof remains open. I have been describing methods – electronic and mechanical – which produce what *looks* like a proof, and might well be more dependable than anything printed on a proofing press. These 'soft' proofing systems are increasingly replacing proofs taken from printing plates. Printers and trade repro houses – especially the latter – have every incentive to avoid the time and cost of reproofing and meet customer demands for recognisable, measurable quality from the outset. Many printers rely exclusively on trade repro houses with equipment which they could not justify installing themselves due to its capital cost and operational overheads.

Yesterday's proof tended to be the starting point for an argument between customer and printer about what was really required; today's should simply confirm that the customer's specifications have been carried out. It is still possible to receive a proof of a colour job produced by a proofing press on 'proofing stock' – high quality coated paper – which bears only an approximate resemblance to what will be produced by the production press on a completely different grade, weight or colour of paper. Such proofs are near to being useless, and whenever possible the print buyer should specify or supply a stock as close as possible to that intended for the production run.

The alternative for many buyers has been to require production press proofs, and to get them they are prepared to go to the printer's plant and pass all important jobs at the press. A production press proof will certainly be closer than one made on a proofing press as an indication of what can be expected from the actual run; but for the printer (and, therefore, also the buyer) it is an expensive business to plate up a press, do the makeready and run it for a single proof. It can only be done for sheet-fed work and even then it cannot show the variables which might affect colour balance and consistency over the production run.

The so-called 'soft' proofing systems are of three kinds: for viewing colour transparencies before scanning; for viewing colour pictures after scanning but before output from the scanner's film separations; for viewing a job after output as a set of combined separations. None of these reproduces colour exactly as it will appear on paper, but simulates colour values. How can a buyer be certain that these simulations are close enough to be of use? He cannot, though he can get some information by directly comparing the 'soft' proof produced on a screen with proof of the same subject printed on a proofing press. This kind of demonstration is not easy to set up and is reliable only for the specific illustration which is the subject of the comparison.

In electronic proofing a distinction must be made between systems which project colour pictures onto a screen, or VDU (which are 'windows' on what has been stored in a computer) and 'hard' proofs from electronic sources. Electronic colour proofing will probably develop rapidly, as do such systems these days; meanwhile the dry toning methods already described are, for the present, more likely to be encountered, and are generally more reliable and easier to use to known proofing standards.

Electronic colour scanners

Colour scanners have been around for some years. Until recently, however, they were strictly for making colour separations electronically. This many of them still do, and colour scanning has largely replaced conventional camera-based repro in most printing industries; but colour scanners can now do a number of other useful things. Their development has been aided by the electronic pre-press systems already described. As electronic devices the scanners are very much at home with electronic input. The colour scanner replaces the techniques and tools of photo-chemical reproduction and retouching. The buyer who thinks only in terms of conventional camera repro is rapidly getting out-of-date.

Despite the automation of what was, and still is in some places, a highly developed set of traditional repro skills, the scanner operators are by no means automata. The creative use of scanners is as demanding of operator skills as are the manual ones they replace. Those who buy their repro from colour scanners need to come closer to the production side of the job than they did when it could safely be left to traditional repro techniques.

The principle of colour scanning is similar, up to a point, to the scanning used in television transmission and reception: the image is scanned in lines by a scanning head building up a composite image of the colours and tonal gradations of the original. That image is presented to what, on a TV set, would be the receiver. The colour scanner deviates from this analogy by requiring a second scanning of the image which converts tonal values into signals of varying intensity which are stored, in analogue form, as a continuous tone (ie unscreened) picture. This is the 'image in the scanner', stored in the form of data which must be converted from analogue to digital data ('pictures to numbers'). This the scanner's computer stores as a digital record which, by then, includes encoding the dots which make up the screen values of colour or halftone pictures for printing.

What happens next will depend on what is present in the scanned original. The scanner has not only recorded the original, as would a photograph of it, but also modified its values and stored them in a computer.

Put at its simplest, scanners 'collect' information from pictures, type and other graphics and turn them into useable data. That store can, under the operator's control, be manipulated in some quite spectacular ways not available to

COLOUR

the process cameraman, such as creating a wider range of tonal separation in a 'flat' original, or enhanced sharpness in an unsharp one. The camera's 'data' are what appears in front of the lens: it can be changed a little by various tricks of the trade used for in-camera correction, but not much. Manual retouching is almost always necessary.

In electronic colour scanning signal intensities can be modulated more extensively to affect tonal values. The resolution of a scanner is the number of lines an inch it scans to collect the graphic data. Scanners exist which make as many as a million 'observations' (records) per square inch, which might not be necessary for colour pictures but for type or fine-line subjects is essential. The scanner's output can also be modified *after* digital storage: what goes in need not determine what comes out. A computer store is immediately responsive to many more modifications than can be made to the component parts of an image by manual retouching. All this has made scanners familiar and useful – though not invariably well-understood – 'tools of the trade'.

Where next? Scanner development has not been as fast as once predicted but has occurred nevertheless. The earlier type of scanners used a revolving drum, on half of which is mounted the original being scanned by the 'reading' head which transmits the data to the computer. In this arrangement a halftone screen could physically be incorporated on the scanner's drum, but this made it difficult to reproduce lines (which would be broken by the screen ruling) and made the scanners unsuitable for reproducing type. Later scanners use a 'flying spot' which 'writes' dots onto the film from a helium neon laser at the output stage, making for a more sensitive range of controls encompassing fine line work and type.

Then the scanner went further, and became capable of scanning and outputting whole pages of text and graphics. This 'full page' output was achieved by using 'pre-scanners'. These work along similar lines to the computer-assisted design terminals already described. Control is passed back from the scanner operator to a station where much more can be done to edit, modify and assemble the pages before they are finally scanned and output. What might once have been called 'camera-ready artwork' can now be produced as 'scanner-ready input'. Why, it could be asked, should anybody go to the trouble and expense of using electronics while good old Bristol board is still available? The answer is that when *all* required data are present they can be used immediately as scanner input. This has been called 'a new era in colour reproduction', which indeed it is. Instead of the scanner being used as a recording device, operating only on an original, it can accept recorded data from other sources, such as magnetic discs, which have been made 'scanner ready' by operators working on screens with cursors, light pens and other input devices. Keyline artwork can be dispensed with – the computer 'draws' far more accurately than most artists – and pictures can be merged, montaged, enlarged, reduced, cropped and altered in numerous other ways. High magnification allows detail to be added using a technique called 'electronic airbrushing' which adds colour units (the 'pixels' we met before), and defects in the original can be eliminated or corrected, highlights inserted and even whole areas of the picture changed, and special effects introduced.

There are now few snags, apart from the sheer capital cost of the equipment, though proofing remains a problem. Proofing is easy at the origination stages, but once the data is in a computer it does not exist in a visible form, so it is never simply a matter of 'pulling a proof' of the finished work.

Scanner technology permits full colour pages to be transmitted over satellite communications links from one part of the world to another. This is an extension of facsimile transmission, which I shall be discussing in a later section.

10
Presswork

If this were a book about printing rather than print buying the section on presswork would need to be extensive and detailed. As it is not, we can concentrate more on what printing machines produce, and their generic differences, and less on how they work. The principles of the main commercial printing processes – letterpress, offset litho, flexography and gravure – are well-established and, with some exceptions, have not changed for a very long time. They have high mechanical speeds, and print a wide range of products.

The buyer might decide whether a given job is better produced sheet-fed or web-fed offset; or he could have the matter decided for him by what is available, at a given price, to meet his specification. It is possible, also, that the buyer's choice will extend to reproduction equipment such as copying, laser printing, in-plant production or 'soft' media such as microfilm, computer store or electronic printers, all of which are dealt with in this book.

It would be wrong to assume that production can safely be left to look after itself once it has been decided how a job will be originated and processed. However much care and attention has been lavished on the preparation stages, reproduction to defined standards in the required quantity to a given delivery date will call for the print buyer's attention from origination to production.

Today's print buyers differ from their predecessors who worked in a more predictable and settled world of commercial printing. Print buying is no longer a desk job and it is now often necessary to monitor production since it can influence the ways jobs are planned. The best and quickest way of doing so is to go to the printers or trade service and look at the plant, ask questions and get information. Most printers still have more than a trace of the artist-craftsman left in their makeup and this does not encourage them always to be explicit and communicative about what their production machinery does. Print buyers should not hesitate in asking.

The movement towards specialisation of product categories has already been noted. It is likely that specialists will offer inducements to buy their specialisations. Mail order packages, for example, can now be produced on presses which not only print them, but also carry out a large number of extra operations, such as perforating, folding, gluing, inserting and even making and printing the envelopes on the run. Business forms presses are equally versatile in providing in-line options. In-line production of finished print is an important result of the high speeds now attainable by modern, web-fed presses, but the benefits of such speed are secured only if such speed can be attained in other operations. Clearly if operations additional to printing can be carried out at run-of-press speeds this will be possible.

To utilise on-press facilities puts an onus on designers and planners of print. If a strip of adhesive is to be placed in an exact position on a printed item during a production run by the press, that strip is as much part of the design as the graphics. This is also the case when off-press finishing is used, but the latter is more adaptable. An item of print originated for in-line production must *invariably* be considered as a whole, and all folds, perforations, attachments, etc, which will be carried out settled before the job gets to the press.

Another occasion when the buyer may need to leave his desk and work directly with the printer is where pre-press planning systems are available. These computerised systems are really there for automating press control settings, but they directly involve the buyer in that certain controls can be preset. Some printers now have extensive – even lavish – facilities for buyers, designers and others to work 'direct to press' in the plant; the mountain has, at last, come to Mohammed! At the present state of this technology its practical uses can be overstressed. It does not guarantee that, under computer control, the press will behave *exactly* as the program dictates, though it goes a considerable way towards it. Pre-planning systems keep some buyers out of the machine room though!

Print buyers likely to be found *in* the machine room are those with enough experience of what can go wrong on the press to spoil a job. If disasters happen there, expensive materials and capacity have been used to no effect. In the event of a dispute about print quality the buyer should know how to recognise defects caused by careless, unobservant presswork and be capable of arguing any case he has for not accepting delivery of a spoiled job, or part of it. He cannot stand by the press during its production run prompting the operator. If jobs are marred by defects the printer can blame the paper, the inks, the trade service, the designer or the weather and, in some instances, he could be right. If, for example, a printer is expected to use a customer's paper delivered immediately before the job is run, he has little chance of conditioning it before the run. Many print defects can be traced to paper problems. The buyer should, however, be equipped to detect the cause of the commoner causes of spoilage. With fast-running presses some of these can account for a large number of spoiled impressions and, therefore, waste of costly time and materials, before they are seen and corrected by the pressman. They include the following:

Doubling: This is not, as might be supposed, always seen clearly as a double image, which is rare and usually confined only to a few sheets. It is the *dots* in colour printing which double and produce a lack of sharpness and definition in the printed image, as well as upsetting colour balance. It is usually caused by one of the colours (probably the first to be printed) being picked up on the offset blanket of the second printing unit which transfers it to the next sheet to pass through that unit. Doubling can be seen under magnification as a 'ghost' dot.

Slurring: The word is self explanatory. The cause is probably due to the movement of the paper or the printing plate on the press or, sometimes, to over-inking which leaves the image vulnerable to slur in its passage through the press units or the delivery, and even thereafter. Inspection will show whether the slur was caused on or off the press. If many sheets contain slur it was caused on the press and the pressman had not inspected the printed sheets at delivery, detected the slur and quickly corrected it.

Trapping: This affects colour printing when one colour ink is printed over another while the latter is still wet, a necessary condition for the rendering of a wide range of colour combinations. The defect should properly be called *non-trapping*, since it is caused when the first impression does not take up sufficient ink from the second, overlying one, and therefore upsets colour balance. It is the printer's responsibility to use inks which have sufficient 'tack' (miscibility) to combine consistently and well over the run, and to maintain press settings to avoid poor trapping. This, incidentally, is a defect which no computer-controlled press system can correct: it has to be done by the operator.

Tracking: This cause of spoilage can often be avoided at the planning stages of a job, and might have been caused by a lack of forethought at origination. It occurs most often when a sheet contains a mixture of image densities and white (unprinted) spaces. Large areas of solid colour adjacent to small areas of text present the pressman with the problem of deciding how much ink is needed adequately to ink the solids without over-inking the lighter areas, or getting ink into areas which should be clear. Sometimes a compromise is possible, other times not. If not, little can be done on the press to avoid the 'tracking' of the solids into unprinted parts of the sheet, the infilling or over-inking of lighter parts (type or fine-line illustrations) or, if the ink-film is reduced to avoid this, an under-inked solid: the problem was probably potentially there at the design stage.

When planning a job where combinations of solids and lighter areas are present the imposition scheme should be arranged so that, so far as possible, the solids are grouped horizontally across the sheet, not vertically or randomly within it so that a heavily-inked solid 'collides' with a lighter or uninked area in a page or spread. If that is not possible large solids might have to be printed separately from lighter areas. Printers are well aware of the dangers of tracking and will advise on imposition schemes which minimise the risk of it. However, they are unlikely to refuse to accept a job designed and presented so that tracking could occur or, if it does, take the blame for it.

Tinting: Tinting is colour where it is not wanted. It is usually caused by the damping solution used on the press receiving small particles of colour pigments from the inks. When this happens the diluted pigments are conveyed by the damping system to the plate and transferred to the paper. Tinting will be shown mainly in the lighter tones and in unprinted areas. A similar effect can be caused by a scum which forms on the plate (due, possibly, to impurities in, or the bad formulation of, the damping solution) preventing non-image-bearing areas from repelling the ink and thereby leaving marks. Scumming is less likely now that improved damping solutions which include alcohol are used to prevent the build-up of scum on the plate.

Hickies: An American term which covers all kinds of flecks, spots and other small marks on a printed image, particularly in solids. These can arise from many causes, but a genuine hickey is a dark area surrounded by a fine white line or an irregularly-shaped white fleck. The former is usually caused by a fragment of the skin which forms on the top of ink in an opened ink-container finding its way into the inking system of the press. The latter may arise from paper dust or other debris carried onto the plate by the offset blanket. Both can be prevented from being carried through an entire run by the pressman seeing them and stopping the press to clean it. Detection of hickies is harder at high press speeds and on web-fed presses where stopping the press is, in any case, a more time consuming and waste-producing matter. Unfortunately it is high speeds which increase the risk of hickies, as do low-grade papers. Some papers used in offset-litho are extremely 'dusty' and produce a lot of airborne debris, especially at the folder.

Picking: Printing inks require 'tack' (adherence) to 'stick' to the paper, or to an already inked area. Too much tack can damage the paper and even tear bits of it away. This might seem easily avoidable by the printer, since it is he who chooses the inks; but paper quality is a contributing factor, and the choice of paper could be the cause.

Showthrough and strike-through: Some of the worst problems of showthrough may be avoided at the specification stage. The phenomenon is one of paper. If the images on one side of a page show through on the reverse side it is because the paper is not sufficiently opaque to prevent it, and little can be done about it. Showthrough is not the same as strike-through, though its appearance is similar. Strike-through is caused by the paper absorbing ink and *conducting* it through its fibres to the reverse side of the sheet. Over-inking, especially of solids, can result in strike-through, especially on the more absorbent papers. The extent to which a printed image can be spoiled by either showthrough or strike-through varies, but both can be more easily avoided when the paper is selected. Many printers are equipped to conduct tests on paper for this, and other paper-related problems. If they are it is to the buyer's benefit to have them made and take notice of the results.

Dot gain: Dot gain is the increase in the size of the dots due to the compression of the ink-film which makes them larger on the paper than they are on the printing plate. It is therefore something which happens *on the press*, and will not be visible in a proof. The effects of dot gain are a shift of tonal values and a loss of image sharpness. To a small extent it is inevitable, and tolerable, but it should be controllable, especially since the percentage of dot gain is influenced not only by the press, but also by the paper. The unpredictability of dot gain is one of the most intransigent problems in the way of establishing standards of proof comparisons with press runs. Dots can be reduced deliberately during exposure to plate to compensate for a known percentage of dot gain on the press, but this is not always easy: different presses produce varying percentages of dot gain, and this factor must be taken seriously if separations made for sheet-fed offset are used on different papers and on web-fed machines.

Misregister: Gross misregister is easily detected, but a small 'slip' in register which mars crispness of appearance and colour values for a small number of impressions before it is corrected on the press is harder to find. Most modern presses have electronic register controls, but even these do not appear to eliminate the many examples of poor register which can be seen in much long-run commercial colour printing. There is little the buyer can do about it, short of rejecting a job which is so conspicuously out of register that it is completely spoiled. If so only a re-run will remedy the situation, and that is not always feasible.

Misregister will hardly ever be found in proofs – the printer sees to that – but the print buyer cannot examine every impression on a production run and relies on the printer to maintain sensitive registration.

Poor paper quality or conditioning is the commonest cause of misregister, especially on fast-running web-offset presses where the paper is unwound from a reel (which must not be deformed) and passes at some tension through the units and, en route, possibly through a dryer, and can be stretched, or may contract, in its passage from one unit to the next. Paper conditioning is the printer's responsibility, but he cannot be held responsible for conditioning paper if it is supplied by the customer immediately before the run is started.

For in-line production where, in addition to printing, the job is folded, cut, perforated, or has any other operation carried out at run-of-press speeds, registration of print to cuts, folds, etc, is as important as registration of image. It is often possible to make some allowances in the original design to anticipate this: items such as business forms should be designed to allow generously for the registration of perforations, numbering and other elements in addition to printing.

It is easy wrongly to diagnose misregister. Doubling (qv) and dot gain (qv) can superficially resemble misregister to the naked eye. A magnifier should always be used to examine print for such defects, but the buyer must know what he is looking for.

Misregister is less likely on sheet-fed presses, where it can be more quickly detected and corrected by the printer, though the latest sheet-fed machines also run at very high speeds and visual detection can be more difficult unless electronically controlled, or unless frequent checks are made by the operator at the delivery end of the press.

The greatest risk of poor registration in sheet-fed work is when a job has to be run more than once through a press as, for instance, it would if a two or more colour job is printed on a single-colour press. This should be avoided for all but the simplest and lowest quality work.

The cost of materials is now a powerful reason for getting

things right first time. Materials waste has now become an important aspect of print production. As materials prices increase waste margins represent a significant proportion of costs. Waste is likely to be higher for long-run, web-fed work to exacting standards, but it is also a potential expense in any printed job. It can be reasonably suggested that the responsibility for waste control lies mainly with the printer. This is true up to the point where the job has been planned and prepared in ways which minimise the dangers of materials waste which is the print buyer's task, in collaboration with designers and others concerned with specification and origination.

On the machines, and mainly on the presses, some of today's technology is directed mainly, or entirely, towards waste reduction. On-press devices, such as web-scanners, automatic register controls and plate scanners are used to reduce waste at high running speeds, which prevent machine operators from detecting variables in print quality before a large quantity of material has passed through the press. A further cause of wasted paper on reel-fed presses is the fact that if the press has to be stopped for any reason before the completion of a run, the period during which the press is being 'run down' from production speed and after restarting 'run up' to speed again, will inevitably result in a number of impressions being lost because of low print quality. This is one of the disadvantages of the offset litho process: the inking/damping balance has to be restored which takes time, and therefore consumes materials on the press.

A survey on sheet-fed waste[1] alone indicated that the printing industry's average level of spoilage was three per cent, where spoilage is defined as 'a loss caused by an error of judgement and/or performance that results in all or part of a job being unsaleable'.

---

[1] *Waste and spoilage in the printing industry* (including the results of an on-site research study), National Association of Printers and Lithographers, USA, Publication No P111); 1983.

The American study took six years and involved over a thousand companies. Waste margins vary, in sheet-fed litho, between two and nine per cent in the USA.

It is necessary for the print buyer to be aware of the growing importance of waste in terms of cost, and of the measures which can be taken to reduce it. These include waste-conscious job planning, which implies close liaison between buyers and printers at a technical level. It is useless to request an 'estimate' of possible waste before the job is run; no printer could give an accurate one. In theory the printer pays for the waste himself out of his profit margin on the job. In practice he knows that, for certain categories of work, there could be a high waste factor and estimates accordingly. If waste levels are high at a time when materials costs are also high, they are inevitably reflected in total print costs.

There is at present no standard for waste allowance, and the best that a buyer can do, on his own part, is to consult the printer on any aspect of the job which could increase waste, and plan the job so as to avoid it wherever possible. A further step could be for the buyer to satisfy himself that the printer is using up-to-date equipment for waste prevention, and is actively monitoring waste. It is worth asking a printer whether he uses a register system, tests materials such as paper and film, has pre-press equipment which standardises exposure and development (automatic processors) and, in the case of long-run reel-fed work, has an automatic reel splicer on the press, which permits the press to continue running at a consistent speed by splicing the start of a new reel to the end of the exhausted one at the reelstand without slowing or stopping the press, thus maintaining production speed and, most importantly, preventing the waste which results from stopping the press for reel changes.

In addition to materials and print costs there are cyclic shortages of certain materials which directly affect printers' customers, force further rises and temporarily reduce the availability of certain grades of paper and other materials which might be the best, most economical and most appropriate ones for the jobs to be printed.

If the buyer supplies any of the materials – especially paper – it is all the more necessary actively to monitor waste at the press and, wherever possible, to discover how it arises. If it is the result of poor warehousing, bad handling or transportation or other conditions outside the printer's control the printer cannot usually be held responsible for waste and any losses due to it will be paid for by the customer.

Offset lithography is immensely versatile but, despite the level of automation and sensitive controls which are now used, it remains a matter of some skill, experience and knowledge for a printer to get the best out of it. The principle of offset litho reproduction requires not only a critical balance between inking and damping on the plate at the moment the image is transferred to the stock, but also the maintenance of that balance throughout a run during which conditions on the press vary from moment to moment. For the customer it is difficult to identify the cause of print defects with confidence, though a print buyer should have looked closely and critically at enough print to detect the commoner causes of spoilage so far mentioned and decide whether they could have been avoided with more care on somebody's part. Some defects can appear on only a few sheets in a run, while others persist and may spoil a large amount of printed materials.

Papers and inks respond differently in differing combinations, and tests may need to be made on selected substrates before a large or important job is run. The print buyer should either understand what the tests are designed to show, or ask for an explanation. Some organisations which buy large quantities of expensive print set up their own testing labs: the equipment must be used with scientific accuracy. A test strip might, for example, show whether the ink vehicle is being absorbed by the paper leaving a powdery pigment lying on its surface – 'chalking' – which can be avoided by changing the ink, using a less absorbent paper, reducing the press speed or specifying a litho varnish. Once on the press it is too late to do anything about it. Mottling of an image can sometimes be traced to paper which is not of the same degree of porosity all over the sheet. Gloss inks must be matched to gloss papers. Wavy edges to a paper stack can cause slurring.

So many and varied are the possible causes of unsatisfactory print that a print buyer would have to be extremely confident to point to one as being unquestionably the printer's fault. Usually the buck is repeatedly passed! It is, however, possible to arrive at a subjective assessment of a printer by simply visiting his plant, talking to him and to his staff, and noticing whether they have enthusiasm and technical confidence, and can convey it to the customer.

11
Binding and finishing

Most documentation needs binding or finishing of some kind. Certain long-run categories of work, such as newspapers, magazines, catalogues, business forms and similar standard-format products, can be printed on presses which have in-line finishing devices. Much larger and more varied categories of print may involve finishing operations which are complex, more difficult and at least as expensive as putting the image onto the paper. At one time the printer's bindery was the most labour-intensive part of the whole plant and, in some cases, still is. This, for the print buyer, means that all off-press requirements must not only be planned carefully as part of the total job, but also with an eye on what they cost, how long it will take to do them and whether the finishing or binding contributes in some positive way to the use to which the product will be put, and is not superfluous.

Where standards are established for specified products, and catered for in their design, finishing specifications will be easier to define; where they are not they must be related to the ways in which the print will be distributed, handled, stored and used. Print preparation and specification can be regarded as an exercise in industrial design: if it is to go through the mail its total weight affects the cost of postage which, though small in units, rises alarmingly with volume. An unnecessarily high post bill could come from no more than a fractional weight increase, which might have been avoided without harm to the product.

Now that paper and board prices are high many print

buyers look to paper alone as a means of reducing overall production costs, including postal costs, yet neglect to consider binding and finishing as part of the deal. To reduce paper weight can have an adverse effect on the quality of print and the speed of production, while a careful consideration of binding and finishing could reveal new ways of cost-control. A serious, and partially successful, effort at greater standardisation of paper and envelope sizes has helped; but it is not always feasible to standardise. The alternative could be quite a large amount of expensive materials ending up on the cutting room floor.

If a report must be filed the dimensions of the filing system determine whether it can be conveniently accommodated. Other things may need to be anticipated before ink touches paper. Lamination, for example, is sometimes a luxury and at other times a virtual necessity, as it might be for the cover of an operator's manual or parts list which might be handled a lot and exposed to marking by oil and dirt. Colour coding, stepped indexing or spiral binding might be indicated for certain jobs where the book opens flat; but it is easy to tear pages away from a spiral bound book if the paper is not sufficiently strong, and if the binding does not last as long as it should the rest of the book disintegrates. A parts list which is likely to need fairly frequent amendment and updating could be more functionally designed using a comparatively expensive loose-leaf binder, such as a ring binder, so that those pages which get out of date can be removed and new pages inserted. These are just a few examples of factors which must be identified at the planning stage if the print is to 'work' properly as well as look good.

The variety and complexity of many print finishing operations means most printers need to call on trade services for some jobs. Print buyers can also buy directly from trade finishers. They do not want plant bottlenecks or idle machines which take up space and labour and to have to await appropriate work to fill them. But it is always prudent to tell the printer how the *whole* job is designed, even if he is only carrying out part of it. The decision to laminate or varnish, for example, could well determine what ink formulations and colours can be used for printing. Some ink formulae react adversely to varnish and to the heat which bonds the transparent cellulose acetate or polypropylene laminate to the stock. Varnish can, in some cases be applied as part of the press run. If so the printer should know what inks to use to avoid spoilage.

Imposition schemes must be taken into account for certain finishing operations so that a sufficiently wide paper margin is left for the grippers of the finishing machinery to work on the sheets without interfering with printed areas. Such practicalities must be the concern of someone, be it the designer or the print buyer, if a well-printed job is not to be compromised at the finishing stage, or made unnecessarily expensive by having to be adapted to the mechanical or size limitations of the finishing machinery.

The wider product scope of in-plant printrooms has encouraged manufacturers to make a number of small, simple, easily operated in-house bindery and finishing systems for documents, lamination and even more elaborate post-press needs. Provided the throughput of work which can economically and efficiently be handled is sufficient to justify their cost and the space they take up, their benefits in-house are not difficult to evaluate; but it takes longer to operate bindery equipment of this sort than to produce the actual print in-plant, and the introduction of binding and finishing devices can interfere with the scheduling of in-plant jobs. If additional personnel is needed to cope with binding or finishing, straightforward cost comparisons between producing work in-house and sending it out to a trade binder or finisher could indicate which is the more sensible course. If the in-plant printroom manager is expected to cope with printing, binding and finishing he may do so at the expense of the throughput of his plant.

Envelopes

Envelopes call for special attention and, though they are more in the nature of supplies, may also be considered in a print-finishing context. Give or take a few thousand the UK post office handles some thirty-seven million postal items (excluding parcels) a year. Most will be in envelopes, some of which will be printed. Around fourteen million envelopes a year are made in Britain. An envelope can cost a lot or a little, depending on its size, the quantity ordered, its quality and whether or not it is printed. If printed envelopes feature in a print job they should be costed as part of the total cost of the job.

An envelope is a really quite elaborate piece of paper converting. The basic questions the print buyer might ask are: (a) Is it better to have it made into the shape determined by the print, or should it be the other way round – ensuring that the print is designed to fit the envelope? (b) Will it do the job? That is, will it not only protect its contents, but also suit the print which it will contain, such as having its window in the right position for a printed-out address to be seen? (c) What benefits, if any, can be expected from bulk buying of fixed-size ranges of envelopes which will, of course, influence the dimensions and maybe the design of the print they are to enclose?

The UK envelope manufacturing industry can be broadly divided into two parts, one producing stock envelopes and the other specially-made, or 'bespoke' envelopes. A few suppliers make both sorts. Obviously, stock envelopes ordered from a manufacturer or merchant will be supplied at a lower price than ones which must be made up specially.

A long-run job for mailing could benefit from the speed and efficiency of automatic envelope-filling machines which (depending on contents) can fill between six thousand and ten thousand envelopes an hour. To work at these speeds the envelopes themselves must be designed for fault-free running through the fillers. Envelopes with pointed flaps, or those not made to accurate dimensions, cause snags and hold-ups for automatic fillers. If this happens the job schedule could be delayed by the need for hand-filling. These practicalities illustrate how observant the print buyer has to be to avoid the waste of resources which can result from a neglect of internal liaison at the early stages of planning a job and seeing it through.

Part Two

Technological Changes

The following sections deal with systems which have become, or are becoming, part of the preparation and processing of text and graphics, or have specialised printing applications. Their uses will depend on the kind of work which is required. Most are available either from trade services or bureaux. In all cases the buyer must first decide where they can be obtained, what they cost and whether they fit the needs and scheduling of the job. In most cases the job must be designed and planned so that the best use can be made of the service. This usually calls for close liaison during origination to ensure that a job arrives in a form which permits a specific system to be used at optimum efficiency. In some instances (for example ink jet or laser printing) a job cannot be handled at all by the newer systems unless it is planned and scheduled for it from the outset.

Only an outline of the technology – and by no means all that which now is, or could soon become, available – is given. Capabilities differ, and it will be necessary for the print buyer to discover *precisely* what a chosen system can and cannot do before using it, and to see samples of its output. Time must be allocated for any additional preparation and if print has to be transported, finished or semi-finished, from one place to another.

12

Copiers

Office copiers have been around for long enough to have had limited, but fairly obvious, in-house benefits for the print buyer or printroom manager. They largely replaced spirit duplicators and, for facsimile reproduction, have been useful for short run work of a kind which would have been too expensive to print. Often, however, the copier has been poorly costed.

Over the years copiers have become more versatile and faster and the more advanced ones have moved quite firmly into the 'reprographic' field. With such add-ons as collators and automatic document feeds, copiers can now provide a viable alternative to small offset printing machines and, with the availability of word processors which produce clear, corrected and edited originals for reproduction, their uses have expanded.

It may, for example, be desirable to use copiers for producing 'hard' information which needs frequent changes and revisions, such as address labels for regular mailings or price lists, where the originals for copying can be printed out by word processors or as computer printouts. These hold the store of data compactly, and can rapidly retrieve it, amend it and provide the revised text for a copying run, replacing more elaborate systems which do similar jobs, but in different ways.

Copying speeds of conventional desk-top or small floor-standing machines are between five and thirty or more copies a minute, with enlargement or reduction of originals. The image quality is good for text, line, solids and even halftones. One has a paper tray which takes 2,000 sheets, an indication of the run-lengths expected from copiers now. Machines which will copy in colour have recently become available. Clearly copier manufacturers see their products as directly competitive with small offset presses for many categories of work. At one time these would have been an automatic choice for copying runs of over fifty. Now numerous 'instant', or 'on demand' printshops offer copying services, and some sell short-run colour copying which could not be handled economically by printing presses.

The latest generation of copiers has increased the challenge to conventional printed reproduction for suitable short-run work. One of these is a digital/laser copier comprising a 'reader' and a 'printer'. The reader works rather like a conventional copier, but the image it scans is focused onto photodetectors which convert it into a pattern of electrical signals, reconstructed by the 'printer' as a visible image on paper.

Because there are two units one 'reader' can feed four 'printers' and produce up to 180 copies a minute. A configuration of this kind is costly and could hardly be called an office copier, since it would need some seventy thousand copies a month throughput to be economical to install, but such copiers will almost certainly appear in bureaux and high-street printshops.

The latest copiers also have 'intelligent' functions, including a text-editing facility, and it is likely that soon they will be linked directly with word processors as output devices. If so their challenge will be towards small phototypesetters as well as in-plant printing presses in some sectors, since they can then handle both text origination *and* reproduction.

Copiers of whatever kind should, if their throughput is high, be costed as part of the print (or 'reprographic') budget: the biggest risk to budget control is to ignore the overhead costs of using them as short-run machines simply because they have other internal uses which are not so costed. A direct 'per copy' cost can be quickly worked out, though it might also be necessary to make calculations on the less quantifiable basis of what the copiers offer in terms of convenience, time-saving and the capital and overhead costs of other types of equipment and out-services which could be used for the same type of jobs.

The most immediately comparable method to xerographic copiers of obtaining copies of an original is to use it to make a printing plate and print it – or have it printed – on an offset litho press. This could, depending on the length of the run and the availability of the press, be the more satisfactory and economical method of working, though time and materials costs would also need to come into the calculation when comparing it with fast copying machines.

13

Ink jet printing

Jet printers offered the first opportunity to place the actual printing of images under computer control to insert variable text into printed matter, as can electronic and laser printers. Their limitations are the size of the area they can print and the relatively poor typographical quality of the printed characters. They do, however, have the considerable advantage of being capable of physical incorporation into a printing press so, like a numbering box, they become part of a production run, though mechanically independent of it.

Most often they are used for bulk addressing. Their other advantage is an ability to print on formed (ie not flat) surfaces: oranges and eggs are among the items which have been printed by ink jet. They have also found applications in continuous stationery production for numbering, dating and other variables within a standard form.

Jet printers use water or alcohol-soluble inks stored in a control unit. The 'writing head' is connected to a control unit by two tubes, one of which contains electrical control leads and the other ink. The jet of ink is forced through a system of nozzles. The modulated jet is electrostatically charged in a metal 'tunnel' under high voltage which causes the ink stream to be broken up into very small, separate drops. A deflecting unit then directs the stream of ink drops in a pattern defined by the voltage charge. Changes in voltage change the direction of the line of drops and 'writes' on whatever is passing under them. The technology is attractive, and can be compared with television insofar as the ink drops behave like the electrons

INK JET PRINTING

which make up an image on a television screen. Each mark is made within a matrix. The electronic part of the system converts the data fed in code from a magnetic store which instructs the charging tunnel what to write.

Ink jet printers use sets of standard characters which have a similar appearance to those of the dot matrix printers used for some kinds of computer printout. They have been successfully used on presses to address newspapers or magazines for postal distribution directly onto the product, thus avoiding the need for wrapping, for printing adhesive address labels and for price and other markings on shelved products.

Their uses are at present restricted but their potential for application is high. The ink drops are generated at as many as sixty-four thousand a second, giving an inscribing rate of over fifteen thousand characters a second, or up to a thousand feet of paper a minute at seven characters to the inch, each formed from a 7×5 dot matrix. The printers are not dependent on the running speed of the press, if they are incorporated in it, and a number of 'writing' heads can be used, one for each line of characters and each with its own programmed microprocessor. There is little or no ink waste since ink is recirculated through the system.

14

Electronic printers

Electronic printers can be seen either as components (output stations) of the integrated systems outlined in the section on electronic publishing (see chapter 19) or as independent printing units. Either way, now costing between £140,000 to £200,000 for the fastest and most versatile machines, they are hardly likely to be bought on a whim!

The capital cost of electronic printers must, however, be considered alongside the economy of operation, which does away with some of their high overheads associated with printing machines – plates, inks, labour costs, handling costs, etc. Electronic printing is the only really new method of printing to have been introduced in recent years.

A computer costs another £20,000 or so, though the actual printing machine can be used without computer control. Programmed control makes an electronic printer the output part of an electronic publishing system. The printer can receive graphics in two ways, one via a scanner which has a range of controls similar to a colour scanner (described on pages 71 to 74) for scaling, reducing, enlarging and contrast modification, but without colour capabilities; the other method extracts digitised images from a computer.

Images are produced by selectively charging the paper and toning it to create visible characters and graphics as does a xerographic copier. What we have, in effect, is a substantial upgrading of the copier which extends its uses in the 'before' (origination) and 'after' (reproduction) areas. The printer itself is essentially a production machine. Compared with even a fast copier it is more versatile and faster. Copiers can reproduce only documents which already exist, and their economic run lengths are still comparatively low. Electronic printers will print on both sides of a sheet, horizontally or vertically, will handle a variety of stock from around 60 to 200gsm and have speeds up to a hundred and twenty pages a minute, though around seventy pages a minute is at present more realistic. Since each page is printed sequentially, collating is no problem.

The essential difference between electronic printers and copiers or printing machines is that the image is formed of tiny dots – ninety thousand to the square inch, as is needed for type and line-work of high resolution. Electronic printers are not, at present, capable of yielding the image quality of offset litho printing.

Bureau services are available for electronic printing applications. The extremely high output of such printers demands continuous (twenty-four hours a day) use by the bureaux to justify the high initial cost. (One calculation by an electronic printer manufacturer is that a level around seven hundred thousand prints a month is needed for profitable use.) Applications have, nevertheless, not been hard to find. Direct mail houses use them for the 'personalisation' of packages from computer-held lists: when you receive a direct mail package addressing you by name the name has probably been inserted by an electronic printer, which can also send you a 'personal' handwritten letter. A large multiple store chain uses electronic printers constantly to update its price-lists. Labels for price-coding products can be produced by electronic printers, and insurance policies are handled by them. Even a book has been printed electronically. It was written by the author on a word processor, the typescript transmitted from the w/p disc to a mainframe computer, the typography, pagination, etc, inserted and the text output to an electronic printer. Electronic bookprinting allows books to appear on the market very quickly, which, for highly topical subjects can be essential for a book's sales.

One advantage which might help to keep electronic printers working profitably is that, unlike conventional printing and copying, they are not materially affected by run lengths. Run length will often decide how a piece of print can be produced. The costs of time and materials for pre-press and the setting up of the press are the same for short or long runs and, therefore, the longer the run the lower the unit cost of each impression. Electronic printers will, provided with the right sort of input, print a few documents as easily and economically as they can print large numbers, and do not demand post-press collating.

15

Laser printers

Laser printers can be described more succinctly if the preceding section on electronic printers has been read and understood: they do a similar job, but more simply and cheaply. They are concerned mainly with small areas of print, such as address labels and the personalisation of direct mail. Most circulation lists for magazines, direct mail and other listed data are computerised and laser printers can be driven by these data bases held on disc files. They will also print letters to reasonably high quality standards, with personalised inserts, at up to ten thousand an hour.

At present there are two sorts, the only difference between them being that one can handle a wider range of typestyles and formats, but will print on only one side of the paper, and the other, while less flexible, will print on either, or both, sides of the paper.

If a print buyer requires laser printing he will almost certainly obtain it from a bureau unless his organisation has a need for the more-or-less continuous use of such machines, in which case he will already know the ins and outs of them.

The cost of laser printers is high (currently £250,000 for one make) and their use, either in-house or as a trade service, demands long runs – at least two thousand letters, and preferably more. This could change now that laser printers which will print on both sides of a sheet are becoming available. Formerly – and still in many bureaux – the printers need a continuous reel, or web, of paper. Laser printers do not, at present, normally offer more than one colour, though a

Japanese laser printer for two-colour printing is on the market.

The 'updating' facility is important but the customer has to do the work which ensures that the input is correctly updated on the mainframe computer before printing. This responsibility for 'cleaning' lists is unavoidable, and a planning period of between four and six weeks is required for this, and for a thorough checkout of the chosen system's requirements. There are some twenty-eight different models of laser printers available at present world-wide.

Papers are also important to successful laser printing (usually they are between 80 and 90gsm) and designs which include folds, gluing, perforations and other features must take carefully into account where the laser-print will appear. No sensible print buyer would accept a request for laserprinting without first checking carefully with the bureau to find out exactly what equipment will be used and what limitations it places on the design and positioning of the laser printed area(s).

Anyone curious about the technology would be well-advised to go to a bureau and see it in operation. It resembles xerographic copying in that it uses a toner to fuse (with a laser) the images onto parts of the paper which have been sensitised to receive them. The toners need the laser because, at the fast speeds of the process, a high temperature (around 200°C) must be quickly and accurately applied to fix the image.

Laser printing is at present a limited, but useful, process which, given the careful planning it requires, will continue to find plenty of applications. Eventually it might take a modest place among the 'new printing processes', though it cannot compete for image quality with electronic printers, copiers and printing presses.

A checklist of basic requirements which the buyer must be satisfied are met before using laser printers would include the following:

☐ Pre-printed stationery must already be available for laser

LASER PRINTERS

printing, and be of the correct size and design for the equipment used.

☐ Some laser printers will print sideways, allowing a saving of paper. Need this be done?

☐ The paper must be of the right weight (between 80 to 90gsm or, for some equipment, from 70 to 75gsm) and it should be delivered for laser printing several days before the actual run so that it will acclimatise to the ambient conditions of the place where it will be printed.

☐ Some laser printers cannot print on both sides of the paper. In this case the reverse side must already be pre-printed. It might be necessary to check that the typeface on the pre-printed side compares closely to that of the laser-printed side.

☐ Features such as perforations, gummed strips, attachments, etc, on mailing packages must be carefully positioned at the design stage so that the laser printer places the image correctly in relation to the overall design.

☐ Check any computerised list being used as primary input for a laser printed run of address labels: no changes are possible during or after the run.

☐ Decide whether the length of the run is sufficient to justify the extra trouble and expense involved, and additional work with clients, designers and producers all along the line. Currently a run of ten thousand is about the minimum at which laser printing can be considered an economic proposition, but this could decrease as the equipment itself becomes more versatile.

☐ At the outset of the job examine the equipment to be used. One system has a dot density of 90,000 a square inch compared with 26,000 for another. The range of typefaces available will differ from one supplier to another.

16

Facsimile transmission

A job produced in different locations may have to be examined at several stages of its progress, often by more than one person. This applies especially to typeset proofs and full pages. When most print services could be bought under one roof the problem was easy to solve: proofs could be sent by the printer, corrected and sent back to him with a reasonable assurance that he knew exactly what to do next. Now that a buyer could be getting typesetting and origination from one place, colour separation from another, and binding or finishing from a third the need for synchronisation is paramount if schedules are to be adhered to.

Equipment for transmitting and receiving line, text and tone has been around for a long time and used for various purposes, such as the transmission of weather chart data from outlying weather stations, pictures (and, more recently, whole pages) for newspapers to a central production plant from distant locations and sending financial information.

Facsimile transmission systems are also used by printers and their customers, and the value of these systems has been increased by the improvements made in communications technology. Data can now be 'compressed' so that actual transmission times are extremely short: at the receiving end it is 'expanded' for printing out. Normal telephone lines, radio and communications satellites have provided an impressive set of options for conveying text and graphic information from one place to another. Not all problems have been solved. Radio transmission is still hampered by the earth's curve and the inability of radio waves to conform to it; line transmission over long distances can suffer from 'noise' which affects image quality.

The equipment most generally used when documents are sent from one place to another is facsimile transmitters working over landlines. There are at present more than sixteen thousand users of facsimile transmitters and receivers in the UK and around sixty thousand worldwide. The difference between earlier and later systems lies mainly in their speed and reliability. For the slowest machines an A4 sheet is scanned and transmitted to a receiver, which prints it out simultaneously line by line in about six minutes. More elaborate machines, using analogue technology, will reproduce a scale of tones from black to grey with a fidelity sufficient to present a tolerably exact representation of an original at the receiving end. By encoding and compressing the image the time is reduced to around thirty-five seconds a page, and the machines are designed to monitor quality and compensate for irregularities introduced on the line as 'noise' during transmission.

Printers have, on the whole, been slow to grasp the benefits of facsimile transmission except where, as in newspapers, these are clearly demonstrable. It can, however, confidently be predicted that machines will soon be available which will transmit an A4 sheet in four seconds or less, and if the machines are leased from the manufacturers, British Telecom will have to make its system operational: it is anxious to do so, since transmission charges are geared to the amount of use the machines get. As the costs of buying or leasing the machines go down and their transmitting speeds go up, more are being installed.

The next stage could be the transmission of text and illustrations together in black and white or even in colour. This is already possible, using very expensive and advanced technology based on colour scanners, and in use by international publishers for sending pages from one part of the world to another for international editions. I do not foresee much practical value for facsimile colour transmission since it is unlikely to be sufficiently closely related to the colour standards used in the printing industry and, of course, it would be exceedingly difficult to incorporate screening in a transmitted image.

Without going a step beyond the already available technology a print buyer could call up 'instant proofs' of reasonable quality on his desk from printers in the UK or abroad. The buyer would be neglecting a potentially useful extension of this facility if he did not also use facsimiles to assist designers in scaling, copyfitting and other tasks which could be delayed by the inaccessibility of rough proofs to scale to speed up the job.

A further development of facsimile transmission is to put a whole platemaking system on the receiving end, so that incoming images, scanned at the transmitter, are reconstructed at the receiver and exposed to a litho plate ready for press. At the receiving end the pasted up artwork is 'read' in a scanner by a laser optical system, fed automatically or manually into an exposure unit which registers it and laser-records it for passing to the processor which produces printing plates at the rate of up to forty an hour. The input can be a paste-up or a page proof; the output can be a litho plate, an electrostatic plate, a relief plate (for letterpress or flexo printing) a paper positive or a film negative.

17

Raster image processing

Many present-day systems, and those which are seen as the most likely developments for the future, create 'networks' in which separate operations such as typesetting, layout and page makeup are either gathered into a single, flexible system in one place, or used as such over a distance.

The concept is that of an 'all purpose' graphics/typographics terminal which not only produces complete pages but can also take those pages directly to 'hard' output, either using electronic printers (see chapter 19), or conventional offset litho printing by working directly on printing plates. In the latter the plates are created not by photographic methods, as is now usual, but from a stored 'page' of digitised data which is used to control a laser platemaker. It goes, untouched by hand, from computer-stored data to printing plate.

It is likely that, in future, the separation of pre-press areas (such as phototypesetting, colour scanning, computer assisted design etc), will be replaced by single, versatile systems which can handle everything which needs to be reproduced graphically. What we now have is a lot of output from various systems which has, somehow, to be brought together in one place where it will be reproduced. What we are heading for is integration which will bring the input together and store it all in one central digital computer to which a number of different terminals are able to work, so that a fully paginated and ready-to-print output is available. It is 'putting the jigsaw together'.

Once you have images stored as digital data, you are a long way towards RIP technology. An RIP set-up might typically

RASTER IMAGE PROCESSING

have a central computer serving several sub-systems: a typographical one, with access to a 'type library' (also digitally stored); a graphics one, using the electronic graphics origination technology described in the section on 'handling graphics' (page 52); a make-up terminal which gathers in all the computerised prepared data and makes it up into pages; and an output device. It is only the output device which is really new.

A raster image processor accepts all data in page form and outputs it on what, for want of a better term, can be called an 'image setter' – a 'picture' of the page with a high resolution (1,000 to 1,400 lines an inch). This can be used to make a full-page proof or can be directed to other equipment, such as photocopiers, electronic printers or laser platemakers for offset litho reproduction. The storage of pixels (picture elements, see p56) is no more complicated than the storage of digitised type: they are both treated by the computer as images.

RIP is, at present, developed to the stage where it will store only black and white text, graphics and halftones; it will not be long before the need, and therefore the equipment, will be available for colour reproduction as well as monochrome text and graphics.

For the buyer the main question will still be: 'Can I get what I want from it at a price I can afford to pay for its advertised benefits?' It is one he should be equipped to ask when it arrives, which might not be long.

The part of this system which the print buyer is most likely to encounter is the preview terminal, now widely in use. It is similar to the screen on which editing and other operations can be viewed at a word processor terminal. Instead of showing the data piece-by-piece for editing, the preview terminal is purely a means of making a final check on what has been stored. It is a 'window' or a 'snapshot' of what the page will look like.

Though some preview terminals are capable of showing text, even to the extent of reproducing typeforms, others merely indicate 'copy blocks', illustration areas and other page-design elements, such as rules and headlines. From this any formatting errors which would prevent the page from fitting precisely (such as columns which run over their allotted length or captions too long for the available space under a picture) can immediately be seen.

After previewing, any further corrections are made at an editing terminal. The value of the preview terminal is that, by presenting a more-or-less exact representation of what is on a terminal's disc store, errors can be corrected before the disc goes forward for typesetting. Preview terminals, in their present form, are really of greater value to the printer than the print buyer, but if the latter knows they are there, there are good reasons for gaining access to them to examine output for errors which have been introduced at make-up stages.

18
Microrecords

There is something rather comforting about microfilm. It has been known, used and has demonstrated its value for some time, and its technology has not changed much. It stores photographed records in reduced forms (one being the microfiche). Microrecords of this kind might not be the direct concern of the print buyer until something needs to be printed from them, otherwise they are simply accessed by viewing equipment. The computer has, in some areas, replaced microrecords, not always with obvious benefit but 'because it is there'.

It is possible to use a microfilm image to make an offset plate, provided the original is sharp and undamaged. Systems have been designed to allow 35mm microrecords to be accessed, sequenced, projected and enlarged under computer or magnetic card control, and exposed onto printing plates as a continuous operation. The user can go straight from an original document to microphotography and from plate processing to printing with fewer steps than would normally be required, but the flexibility of the system is strictly limited to what can be photographed in black and white. To be fully utilised, it needs microrecords in quantity which need regular printing.

The advantages of microfilm-to-plate systems are simplicity of handling, the facility to use various size images and, by reduction or enlargement, fit them into fixed formats and differing imposition schemes. Easy reprinting is also a consideration. This is attained by using 'control data' – a computer or magnetic card reader which contains all the original instructions for printing routine jobs.

Such systems include a projector to transfer the image onto the loaded offset plate and a step-and-repeat machine which will move, position and reposition the microfilm images precisely on the offset plate. Before a plate is made a viewer is available for checking the microfilm, or for making proofs photographically. Specially manufactured offset plates are needed for this little-known process.

19

Electronic publishing

The precondition of electronic publishing is the existence of a database; that is to say data already held on a computer and accessible by a terminal. The amount of scientific, legal and other data thus available increases daily. The value of the computerised database lies almost invariably in the speed at which it can be amended and updated – much faster, of course, than printed data, because it exists in electronic form – but it can, therefore, only be accessed electronically (eg by a computer terminal). 'Publishing', in this context, is concerned with any data (text or graphics) which can be extracted from computers and converted into print. The database may be managed as a separate, commercial undertaking. If so, the 'publisher' must pay for the data it contains (or use it under licence from the owner), then find some means of converting it into a form which allows it to be processed as print.

This might not be easy. Incompatibility between computers and phototypesetters has already been mentioned, and to 'milk' a database for printed output requires an interface which will allow the data to be transferred (made 'portable') from the one medium to the other.

If a database is created with this end in view it is likely to be less complicated and expensive to produce print from it. One fully electronic system lists seventeen product categories which can be stored in a central computer and maintained as data bases for extracting as direct input to phototypesetters which are part of the system, whenever they are needed in print. These products are typical of the kind which, at present, are likely to be considered part of electronic publishing: technical manuals, directories, parts catalogues, textbooks, abstracts, timetables, price-lists, product catalogues, insurance policies, classified advertisements, direct mail, business graphics, census tables, market research documents, Yellow Pages and fares schedules. All have in common the need for rapid, accurate and frequent updating, and all benefit from frequent printed publication. The volume and variety of computer-based data is spreading widely and includes data which might already be held on an organisation's own mainframe computer, or one to which it has direct access.

Though a database is likely to be a mainframe computer, data can also be originated and stored by word-processors, and graphics by scanners or computerised design terminals (such as those described on pp 52 to 58). However the data or graphics are held, they must be managed in a way which makes them available for the kind of output wanted, including print. Many print buyers are becoming more and more involved in working with technologists in other areas, and a basic understanding of what is possible and what might, with money and patience, be made possible, is desirable from several points of view, not the least being that computer technologists cannot be assumed to know in detail what is ultimately needed to appear as print.

For example, if the contents of a database are to be printed a program must be written which does far more than display the data on a VDU. It must also be able to handle text or graphics so as to generate such elements as page-sizes, formats, blank areas for illustrations or advertisements, justification, tabulation, and so on.

It should be added that the output from a database could, with appropriate programming, be interfaced for use with microrecords (see chapter 18) or laser printers (see chapter 15).

Electronic publishing is possible only with the most careful initial analysis and planning of the whole operation which, if printed output is to be a part, the print buyer will need to understand in detail. Systems differ widely but have in common the need for programming designed specifically for their chosen tasks, and the 'portability' of the data or graphics from one system – the print input – to another – the printed output – together with any intermediate proofing/reading stages which might be required. What, unlike other computerised systems we have examined, electronic publishing is not able to do efficiently is to allow for extensive editing *after* the data has been accessed for print: what you see is what you get!

20

Beyond the 80s

Technology forecasts for the printing and communications industries have not, on the whole, proved very reliable. Printing already had an impressive technology before such methods as electronics and data processing offered radically different ways of using production machinery, equipment and materials. Printing looked due for a shakeup in the 60s and 70s but not, it turned out, precisely in the ways envisaged. The potential for change has been rapid but the industry and its customers are conservative: today's 'breakthrough' can still be tomorrow's failure. Technological changes usually offer new or better ways of doing some things for some customers and less, or nothing at all, for others.

Nevertheless, important and permanent departures from traditional, or conventional, methods have taken place (phototypesetting, electronic colour separation, direct word processor input and in-line on-press finishing for example). To succeed, the 'new' technology has to integrate with the 'old', and be capable of working alongside it. We should, therefore, look to a future in which strong existing trends will continue rather than for any dramatic break with the past. The following trends have either emerged or been confirmed in the 80s and I think they will continue to gain ground:

☐ More customer origination of text and graphics via computer links with printers and by direct input from word processors.

☐ More printers concentrating entirely on presswork – making the required number of impressions – and depending

BEYOND THE 80S

on outside services or customer input for many pre-press and post-press operations.

☐ The displacement of small conventional printing plants by in-plant printing or 'instant' printshops using laser copiers and other non-impact systems for some categories of work.

☐ Increased use of communications technology to originate and progress print for production in geographically distant locations.

☐ More plant specialisation, making it harder for the print buyer to find the right printer for the job for some categories of work.

☐ More use of computer-based input by printers' customers, interfacing with the industry's own production equipment.

☐ Fewer printers serving big, specialised markets, but more efficient well-equipped local printers holding their own in high quality general printing.

☐ More sharing of pre-press work via communications networks offering, for example, computerised design, proofing, phototypesetting, full page makeup, colour separations and other pre-press requirements as integrated services 'on tap' by the customer in his own premises.

It is likely in the future that printing and communications will become less preoccupied with developing technology and more concerned with the evaluation, both by the industry and its customers, of what's already available, in everyday commercial terms and for specific applications. For the print buyer this will mean being able to deal effectively with many sources of supply and ensure that all of them do their part of the job within a well-planned schedule to specified standards.

The real value of technological developments will depend on how soon the technology finds profitable applications. The future of printing – and, therefore, the patterns of print buying – will continue to be decided by market forces: what the printers' customers want and what they are prepared to pay to obtain it.

In some cases this has already been made clear. Letterpress

is no longer a mainstream printing process; phototypesetting has replaced metal setting for all but a small part of the industry; colour is in high demand; electronic colour separation has virtually replaced conventional processing, and so on. The industry has, in the meantime, diversified, specialised and become international for some categories of work, but not by any means all.

The message for print buyers is that most print jobs must now be measured very precisely to the means available for their production, planning and preparation. In exchange for these disciplines the buyer reaps benefits in speed, quality, delivery and other important requirements; but only if he organises and manages his print needs in such a way that they fit closely into the technological resources. This is not so novel as it sounds: it has, indeed, always been so. Printing had an impressive and complex 'technology' before anybody used, or foresaw, such developments as computerised phototypesetting, electronic colour separation, high-speed copiers and electronic printers.

The future of printing *per se* is constantly and widely discussed, but it is essential that the situation is seen as a whole. For a start 'print' needs to be re-defined to include other kinds of graphic reproduction equipment not normally used in printing plants, such as high-speed copying, and to take account of the fact that customers may be involved directly in origination and production outside, and independently of, the printing plants.

The printed word is seen by some to be in competition with non-print-based media, such as computer databases, video, teletext and television. So, in some areas, it is. But the question is not whether print will survive these challenges from alternative media – which are really an expansion of communications and not simply substitutes for print – but whether it can identify and serve the new markets created by them. Print is still the cheapest, most efficient, most versatile, easily accessible and, in many cases, the only suitable medium for a wide and varied spectrum of communications needs.

The smaller printing firms today are different from the jobbing printers which they supplanted. They have not only better and faster production equipment but also better management and marketing tools to find work for it (if they haven't they soon disappear). They exchange technology more readily than small firms used to do, have quick access to trade services and are more productive and less wasteful than the old-style printshops.

Some lessons have still to be learned, and better answers to pressing questions still to be found, by suppliers *and* their customers, and the time available to learn them is getting shorter. But the word *Imprimatur* – 'let it be printed' – still has power and meaning. Long may it be so.

Index

Addressing, 38, 100–101, 103, 107
ASPIC, 49–50

Binding, 43, 87–91, *see also* Finishing
Book printing, electronic, 103–4
Briefing, 19, 27, 30–34
British Printing Industries Federation, 33, 49
Budget, 20, 22, 28–34, *see also* Costs
Bureaux, *see* Trade services
Business forms, 35, 54, 55, 82

Chalking, 85
Colour, 61–74
  graphics, 52–8
  previewing, 56
  proofing, 68–71
  reproduction, 64–5
  separation, 62
  standards, 53
  swatches, 66
  see also Computerised graphics, Electronic colour scanners, Presswork
Computer databases, 116–17
  estimating, 30
  graphics, 52–8
  makeup, 56

origination, 58–9
  see also Editing, Electronic colour scanners, Microrecords, Proofing, Word processing
Computerised graphics, 52–8
Continuous tone, 72
Copiers, 29–30, 43, 44, 97–9
  colour, 98
  digital/laser, 98, 103
Copying, *see* Copiers
Corrections, 33
  see also Editing
Costing, *see* Costs
Costs, 25, 84
  control, 29
  design, 41
  see also Budget, Estimates, In-plant printing

Databases, *see* Computer databases
Deadlines, 19
Densitometers, 65–6
Designers, 20, 39–41, 57, 76, 83
  see also Briefing, Computerised graphics, Laser printers
Despatching, 38
Digitisation, 54, 55–6, 98, 102, 111
Direct input, *see* Typesetting,

123

INDEX

Word processing
Dot gain, 81
Doubling, 77

Editing, 53, 55, 113, 118
Electronic colour scanners, 16, 40, 56, 63, 71–4
Electronic printers, 102–4
Electronic publishing, 116–18
Envelopes, 21, 37–8, 76
  filling, 90–91
  standards for, 88
Estimates, 22, 23, 30–31, 32, 36–7

Facsimile transmission, 74, 108–10
Finishing, 25
  in-line, 76, 87–91
  see also Binding
Flexography, 37
Full page makeup, 55, 58–60, 73–4, 111–13

Graphics workstations, 55, 111

Hickies, 79

Illustrations, 52
  specification of, 24
Imposition, 79, 89, 114
In-house printing, *see* In-plant printing
In-plant printing, 29, 42–51
  binding in, 89
  costs, 43, 44, 49
  scheduling, 89
Inks, 66, 85, 86, 89, 100–101
  see also Presswork
Ink jet printing, *see* Jet printers
Insurance, 33

Jet-printers, 100–101

Laminating, 88, 89
Laser printers, 105–7
  designing for, 106, 107
Lithography, *see* Offset lithography

Materials
  costs, 23, 32
  specification, 24
  supply, 37, 38
  waste, 21, 28, 83–5
Microfiche, *see* Microrecords
Microfilm, *see* Microrecords
Microrecords, 114–15
Misregister, 81
Mottling, 86

Offset lithography, 37, 43, 53, 62–3, 75, 82
  waste in, 83–5, 111
  see also Presswork

Originals, 63
Origination, 13, 19, 33, 44, 76
  graphic, 53–8, 76
  text, 45–51
  see also Designers

Page makeup terminals, 58, 60
Paper, 30, 40, 61–2, 77, 80, 81, 82, 85, 86, 87–8
  for laser printers, 107
Personalisation, 103, 105
Phototypesetting, *see* Typesetting
Picking, 80
Pixels, 56 (footnote)
Planning, 14, 17, 20, 29, 30, 87, 88
  pre-press, 76, 106, 117–18
Platemaking, 43, 59, 62, 114, 115
  laser, 110, 111

INDEX

Pre-press, 34, 71, 76–7
Pre-scanners, 73
Presswork, 75–86
Print
  defects, 71–86
  specimens, 22, 23
  transport of, 19
Print buyers
  role of, 7–8, 13–14
Preview terminals, 53, 55, 68, 112–13
Printers
  contracts 32–3
  design services, 40
  overseas, 31–32
  plant capacity, 31
  see also Estimating
Print industry, 13–14, 18, 35
  future of, 119–22
Proofs, see Proofing
Proofing, 23, 30, 33, 40, 48, 67–8, 74, 108, 112
  see also Facsimile transmission

Quality control, 28
Quotations, see Estimates

Raster image processing, 111–13
Records, 37
Repeat orders, 22–3
Reprinting, 34
Reproduction
  of colour 64–5, 67
  standards, 53
  see also Origination
Reprography, see In-plant printing
Revisions, 20, 32
Run length, 30, 32, 33, 34
  for electronic printers, 104, 107
Run-on, 30

Sale of Goods Act, 32, 33
Scanners, see Electronic colour scanners
Scheduling, 18–21, 24, 31
Screen process, 37
Showthrough, 80
Silk screen, see Screen process
Slurring, 78
Specifying, 21–7, 32, 87
Split runs, 33
Standards
  colour printing, 64, 66
  colour proofing, 67–71
  envelopes, 88
  reproduction, 53
  transparency viewing, 64, 71
Strike-through, 80
Suppliers, 35–8
  contracts, 32–3
  records of, 37
  representatives, 35–6

Technology, 8, 13, 14, 15–17, 44, 45–51, 74, 95, 119–22
Time schedule, 24
Tinting, 79
Tracking, 78
Trade services, 20, 26, 27, 35, 37, 44, 50, 88, 89, 95, 105
Trapping, 78
Typesetting, 15–16, 24, 26, 39, 44, 45–8, 51, 54–5
Typographers, see Designers
Typographical specifications, 24

Varnishing, 89

Warehousing, 21, 30, 34
Waste, see Materials waste
Word processing, 17, 26, 42, 43, 44, 46, 48–51, 103, 112, 117